# Cognitive Dependability Engineering
## Managing Risks in Cyber-Physical-Social Systems under Deep Uncertainty

**Lech A. Bukowski**

Director of the Business Engineering Center
WSB University, Dąbrowa Górnicza, Poland

**CRC Press**
Taylor & Francis Group
Boca Raton London New York

CRC Press is an imprint of the
Taylor & Francis Group, an **informa** business

A SCIENCE PUBLISHERS BOOK

Cover illustration courtesy of Weronika de Bończa Bukowska

First edition published 2023
by CRC Press
6000 Broken Sound Parkway NW, Suite 300, Boca Raton, FL 33487-2742

and by CRC Press
4 Park Square, Milton Park, Abingdon, Oxon, OX14 4RN

© 2023 Taylor & Francis Group, LLC

*CRC Press is an imprint of Taylor & Francis Group, LLC*

*Library of Congress Cataloging-in-Publication Data (applied for)*

ISBN: 978-0-367-89730-7 (hbk)
ISBN: 978-1-032-48617-8 (pbk)
ISBN: 978-1-003-02075-2 (ebk)

DOI: 10.1201/9781003020752

Typeset in Palatino
by Radiant Productions

*To my entire international Family*

# Acknowledgements

This publication has been accomplished as a part of a research project funded by Prof. Henryk Walica Scientific Scholarship Fund. Therefore, I would like to express my gratitude to the WSB University in Dąbrowa Górnicza for supporting my scientific studies.

# Preface

The key issue of the book is the uncertainty associated with the operation and maintenance of homo-centric systems composed of cyber, physical, and social parts, each of which can constitute an independent complete system. Such systems, also known as Cyber-Physical-Social Systems, are subject to disturbances and disruptions from a variable and uncertain environment that are particularly difficult to predict due to their high level of structural, spatial, and temporal complexity. These topics are an area of interest in systems engineering sciences, particularly: Reliability Engineering, Safety Engineering, Security Engineering, as well as Dependability Engineering, in which effective methods of dealing with random events and processes have been developed. The methods are based primarily on the assumption of repeatability of events and the possibility of acquiring the necessary knowledge from experience, which in the case of very complex unique systems with a large spatial extent is not fulfilled. It is because in these cases we are dealing with uncertainty beyond the area of random variation, and enter the area of so-called deep uncertainty, also known as radical uncertainty.

As can be seen from the above, probabilistic models based on statistical data are not feasible for the operation of Cyber-Physical-Social Systems under conditions of deep uncertainty, so there is a need to go outside the limitations of the Systems Engineering and the classical models used primarily in the Reliability Theory. Such opportunities are created by the transition from traditional Systems Engineering to the level of Cognitive Systems Engineering, built on the concept of imperfect knowledge, which allows the use of fully unconventional sources of information, such as expert intuition and other not fully conscious cognitive processes. On these assumptions, a new transdisciplinary concept of Cognitive Dependability Engineering was proposed as a basis for effective, efficient, and safe managing the operation and maintenance of Cyber-Physical-Social Systems under conditions of deep uncertainty.

The main objective of the work is to synthesize existing knowledge in the field of operation of Cyber-Physical-Social Systems subjected to disturbances and disruptions from a variable and radically uncertain environment, and on this background to provide a new concept of ensuring its trustworthy performance. For this purpose, a model for solving complex problems by experts, under deep uncertainty, with the support of Cognitive Digital Twins was developed. On this foundation, a framework was proposed for cognitive dependability- based problem-solving in Cyber-Physical-Social Systems operating under deep uncertainty.

The content of the book is divided into four parts consisting of 16 chapters and an introduction as well as a summary with concluding considerations, index and glossary of key terms used in the book. Part I is devoted to an overview of the systems engineering development; starting with the basics of System Science and Single Systems Engineering, through System of Systems Engineering to Cognitive Systems Engineering. The Cognitive Systems Engineering model was based on the concept of imperfect knowledge acquisition and management. Part II presents the five main concepts of attribute-oriented System Engineering; from Reliability Engineering – derived from the idea of failure-free operation, through Safety Engineering and Security Engineering – based on the concept of effective protection, Resilience Engineering – built on the principle of process continuity, Dependable Computing – designed on the concept of fault-tolerant functioning, all the way to the new concept of Cognitive Dependability Engineering. It is a transdisciplinary concept for ensuring that complex Cyber-Physical-Social Systems perform in a trusted manner, both under normal and abnormal conditions. Part III deals with modelling and simulation the operation of Cyber-Physical-Social Systems in a risky environment. It consists of three chapters covering the following topics: methodology of modelling and simulation used for complex systems; modelling of structures, processes, and disruptions for Cyber-Physical-Social Systems; as well as simulation of Cyber-Physical-Social Systems behaviour in a risky environment. The last, Part IV, deals with the problems of managing risks in Cyber-Physical-Social Systems under deep uncertainty. Uncertainty-oriented decision-making concepts are discussed first, followed by a proposed method of cognitive dependability-based problem-solving in Cyber-Physical-Social Systems operating under deep uncertainty. The concept of a cognitive digital twin was introduced to support the process of solving complex problems by experts, and on this basis a framework for cognitive dependability-based problem-solving in Cyber-Physical-Social Systems operating under deep uncertainty was developed. The possibilities and purposefulness of using this framework have been demonstrated on three practical examples of disasters that have happened in the past and have been thoroughly analysed, and the descriptions of reports from these studies have been made available on the Internet. Chapter 17 describes the lessons that can be drawn from these cases regarding the following disasters: collapse of I-35W Highway

Bridge Minneapolis in Minnesota on 1 August 2007; the Three Mile Island Nuclear Accident on 28 March 1979; and damage from the Great East Japan Earthquake and Tsunami, 11 March 2011.

This book is the result of my 50 years of activity in the field of production and service systems operation, maintenence and management, broadly defined. I would like to mention here the names of those outstanding specialists to whom I owe a better understanding of the subject matter in question, and who are no longer among us. These are Professors Wieslaw Zapalowicz, Marian Warszynski, Jerzy Jaźwinski, Zbigniew Smalko, Friedrich-Wilhelm Griese, and Gerhard Schöne. Thank you for your support and willingness to share your knowledge!

The following typographic conventions are used in the text of the book: terms consisting of single, or several words written in italics are included in the Index and those that require unambiguous interpretation are defined in the Glossary. Literal quotations from foreign literature sources are enclosed in double quotation marks like "...", terms requiring additional explanation are placed in single quotation marks such as '...', while the most important definitions and statements are distinguished by a different type of font.

# Contents

## Part III.  Modelling and Simulation the Operation of Cyber-Physical-Social Systems in a Risky Environment

## Part IV.  Managing Risks in Cyber-Physical-Social Systems Under Deep Uncertainty

# List of Acronyms and Abbreviations

| | |
|---|---|
| A | Activity |
| ABM | Agent-Based Modelling |
| ABS | Agent-Based Simulation |
| AHARP | As High As Reasonably Practicable |
| AIR | Average Individual Risk |
| ALARP | As Low As Reasonably Practicable |
| ANN | Artificial Neural Networks |
| AV | Availability |
| BCM | Business Continuity Management |
| C | Consequences |
| CBA | Cost-Benefit Analysis |
| CBM | Condition-Based Maintenance |
| CBR | Case-Based Reasoning |
| CCA | Cause and Consequences Analysis |
| CCF | Common-Cause Failure |
| CDE | Cognitive Dependability Engineering |
| CDF | Cumulative Distribution Function |
| CF | Critical Function |
| CHL | Check List Analysis |
| CIA Triad | Confidentiality-Integrity-Availability |
| CIP | Continuous Improvement Process |
| CIT | Communication and Information Theory |
| CR | Credibility |
| CSE | Cognitive Systems Engineering |
| CST | Classical Sets Theory |
| CT | Control Theory |
| D | Dependability |
| DE | Disruptive Event |
| DES | Discrete Event Simulation |
| DFD | Data Flow Diagram |
| DR | Data Relevance |

| | |
|---|---|
| DTN | Disruption-Tolerant Network |
| DVV | Data Veracity Value |
| EM | Emergency Management |
| EPC | Event-Driven Process Chain |
| ERD | Entity Relationship Diagram |
| ERM | Enterprise Risk Management |
| ERP | Enterprise Resource Planning |
| ESoS | Engineered System of Systems |
| ETA | Event Tree Analysis |
| FAR | Fatal Accident Rate |
| FFBD | Functional Flow Block Diagram |
| FMEA | Failure Mode and Effects Analysis |
| FMECA | Failure Mode, Effects, and Criticality Analysis |
| FRAM | Functional Resonance Analysis Method |
| FTA | Fault Tree Analysis |
| GIS | Geographic Information System |
| GIT | Generalized Information Theory |
| GSE | General Systems Engineering |
| GTU | Generalized Theory of Uncertainty |
| HAZOP | Hazard and Operability Studies |
| HRA | Human Reliability Assessment |
| HRO | High Reliability Organization |
| ICT | Information and Communication Technologies |
| IK | Imperfect Knowledge |
| ITS | Intelligent Transportation Systems |
| IU | Information Utility |
| JiT | Just in Time |
| K | Knowledge |
| KQ | Knowledge Evaluation Quality |
| LSLIRE | Large Scale, Large Impact, Rare Event |
| LSS | Large-Scale Systems |
| M&S | Modelling and Simulation |
| MAS | Multi-Agent Systems |
| MSC | Meta-System Construct |
| MTBF | Mean Time Between Failures |
| MTTFF | Mean Time To First Failure |
| P | Probability |
| PAM | Pro-Active Maintenance |
| PDF | Probability Distribution Function |
| PHA | Preliminary Hazard Analysis |
| PLL | Potential Loss of Life |
| PM | Predictive Maintenance |
| PRA | Probabilistic Risk Assessment |

| PSA | Process Safety Analysis |
| R | Risk |
| RE | Resilience Engineering |
| REL | Reliability |
| RES | Resilience |
| RIDM | Risk-Informed Decision-Making |
| RoI | Return on Investment |
| RP | Random Process |
| RPN | Risk Priority Number |
| SDM | System Dynamics Modelling |
| SDMS | System Dynamics Modelling and Simulation |
| SE | Systems Engineering |
| SOA | Software Oriented Architecture |
| SOC | Self-Organized Criticality |
| SoS | System of Systems |
| SoSE | System of Systems Engineering |
| SPC | Statistical Process Control |
| STAMP | System-Theoretic Accident Mode and Process |
| STPA | System-Theoretic Process Analysis |
| TAR | Throughput Accounting Ratio |
| TBM | Time-Based Maintenance |
| TV | Threshold Value |
| U | Uncertainty |
| UML | Unified Modelling Language |
| UST | Unified Service Theory |
| V | Vulnerability |
| VaR | Value at Risk |
| VSM | Value Stream Mapping |
| VSM | Viable System Model |

# Glossary

*Activity* – an intentionally designed and implemented action. We divide activities into operations, tasks, and decisions.

*Act of a system* – an event that occurs without external stimuli as a self-determined action caused by the system itself.

*Adaptability* – ability to adapt to changed working conditions (e.g., flexibility, agility, ability to learn).

*Availability* – ability to be in state to perform the required functions under given work conditions. It is described by reliability, maintainability, and maintenance support performance.

*Background knowledge* – the ability to evaluate available information and understand reality in accordance with the current state of knowledge, including a judgement of the strength of this knowledge.

*Cognitive dependability* – the ability of human-centric Cyber-Physical-Social systems to perform trustworthy in a variety of situations, including conditions of deep uncertainty.

*Cognitive process* – any of the mental functions involved in the acquisition, storage, interpretation, manipulation, transformation, and use of knowledge.

*Complexity* – a system feature, conditioned by having many diverse and autonomous but interrelated and interdependent components or parts linked through many interconnections that do not allow the understanding or prediction of the system's behaviour based on each component's behavior.

*Confidentiality* – unavailability to non-enabled persons.

*Consequences* – the effects of an activity with respect to the values defined (such as human life and health, environment, and economic assets), covering the totality of states, events, barriers, and outcomes, and often seen in relation to some reference values (e.g., planned values, objectives, etc.).

*Continuity* – a system capability to deliver products (services) at acceptable predefined performance level under the real work conditions (e.g., despite disruptive events).

*Damage* –  loss of physical assets or resources (e.g., infrastructure).

*Dependability* – a collective term describing the time-related operating quality of a system.

*Dependability measure* – the likelihood to avoid disruptions that are more probable and more severe in consequences than it is acceptable for each possible risky scenario.

*Disruption risk* – potential for realization of unwanted scenario leading to a disruptive event with possibility of negative consequences.

*Disruptive event* – an act of delaying or interrupting the process continuity (e.g., system failure, natural catastrophe, manmade fault).

*Disruptive event consequences* – result of a disruptive event (e.g., damage, harm, impact, severity).

*Disruptive risk metric* – the triplet ($s_i$, $p_i$, $c_i$), where $s_i$ is the i-th unwanted disruption scenario, $p_i$ is the uncertainty measure (e.g., probability or possibility) of that scenario, and $c_i$ is the severity measure (e.g., consequence) of the i-th scenario, for i = 1, 2, 3, .... n.

*Dynamic system* (multi-state system) – a system to which events occur, whose state changes overtime.

*Environment of a system* – a set of elements and their relevant properties, which are not part of the system, because they are outside its borders, but may influence substantially the state of the system.

*Event* – a change in the state of the system or its environment which can initiate the start of a process, interfere with it causing errors and pauses, or end it when the desired outcome is achieved.

*Exposure* – a state in which an object (e.g., a system) is subjected to a risk source.

*Extensibility* – the ability of an open system to add new components, subsystems, or systems, as well as new capabilities to a system.

*Flexibility* – means that a given system, depending on the current needs, can be reconfigured, and modified to changing situations.

*Flows* – relations that consist in the movement of goods (e.g., transport) and information (e.g., communication).

*Function of a system* – generating results in line with the goals and objectives of the system, regardless of the state of the system and its environment.

*Functioning* – understood as a manifestation of rational behaviour, involving the fulfilment of required functions, e.g., the execution of production processes or services. The purpose of an operational system should be to fulfil the expectations of the object of the action (e.g., owner, customer, society).

*Goal of an active system* – the preferred state of a system, guaranteeing the production of a given functions at the outcome within a specified, relatively short period.

*Harm* – physical or psychological injury or damage.

*Hazard* – a risk source where the potential consequences relate to safety application (e.g., harm).

*Impact* – the effect that the consequences have on specified values.

*Integrity* – impossibility of introducing changes into the system by non-enabled persons.

*Integratability* – means that a given system is able to form, coordinate, or incorporate into a larger, functioning or unified whole.

*Interoperability* – the ability of connected, autonomous, flexible coupled and usually heterogeneous systems to cooperate and to exchange flows of data, services, material, and energy to and/or from other systems, while continuing their own way of operation.

*Interchangeability* – means that a given system or a part of it can be replaced with another one without losing the basic system properties.

*Likelihood* – a measure for expressing uncertainty, variation, or beliefs, following the rules of probabilistic, possibilistic, or veristic calculus.

*Maintainability* – ability to be retained in, or restored to a state to perform as required, under given conditions of use and maintenance.

*Maintenance Support Performance* – effectiveness of an organization in respect to maintenance support.

*Modularity* – means that a given system (usually to improve its maintainability) is built of functional blocks, separating the system's capacities into modules.

*Network organization* – a structurally, spatially, and temporally complex process organization, whose elements comprise of independent enterprises, and its properties are emergent in their nature.

*Objective of an active system* – the preferred state of the system, guaranteeing the production of a given function in an ultimately desirable degree in a long period of time (a long-term goal).

*Operating quality* – multidimensional measure of system performance related to its effectiveness, efficiency, and continuity.

*Organization* – separate from its environment, controllable, function-producing, and goal-seeking entirety, whose elements are connected by mutual relations, both physical and intangible.

*Portability* – the ability to be readily moved from one environment to another.

*Process* – a system whose elements are events and actions connected by flow relations. A process is a structured chain of events and actions interconnected by flow relations, the aim of which is to achieve the desired result.

*Purposefulness* – pursuing a certain state which can be defined in a strategic and operational scale.

*Reaction of a system* – an event for which another event, preceding it, that occurs to the same system, or its environment is necessary and sufficient.

*Recoverability* – capacity of a system to recover or restoration from a failure, in the acceptable time and costs limits.

*Reliability* – ability to perform the required functions, without failure, for a given time interval, under given work conditions.

*Replaceability* – the ability of one system, component, or person to take the place of another, especially as a substitute or successor.

*Resilience* – a collective term describing the ability of a system to absorb and withstand the disruption impact, and still continue to deliver products or services at an acceptable predefined performance level.

*Resilience metric* – a collective term described by the three main indicators: absorbability, recoverability, and adaptability.

*Resilience to a disruptive event* – the ability of a system to absorb and withstand the disruption impact, and still continue to deliver products or services at an acceptable predefined performance level, as well as the adapt-capacity to a new work conditions.

*Response of a system* – an event for which another event, preceding it, that occurs to the same system or to its environment is necessary but not sufficient.

*Risk* – the potential of gaining or losing something of value (such as physical health, social status, or financial wealth) resulting as outcome from a given activity, planned, or not planned, taken despite of uncertainty.

*Risk analysis* – systematic process to comprehend the nature of risk and to express the risk, with the available knowledge. It includes risk assessment, risk characterization, risk communication, risk management, and policy relating to risk, in the context of risks to individuals, public, organizations, and to society.

*Risk appetite* – amount and type of risk an individual or an organization is willing to take on risky activities in pursuit of values or interests.

*Risk assessment* – systematic process to comprehend the nature of risk, express and evaluate risk, with the available knowledge.

*Risk aversion* – disliking or avoiding risk.

*Risk awareness* –understanding the risk (e.g., the risk sources, the hazards and threats, and the potential consequences).

*Risk characterization* – a qualitative and/or quantitative description of the risk, i.e., a structured statement of risk usually containing the elements: risk sources, causes, events, consequences, uncertainty measurements and the knowledge that the judgements are based on.

*Risk communication* – exchange or sharing of risk-related data, information, and knowledge between and among different target groups (such as regulators, stakeholders, consumers, media, and public).

*Risk evaluation* – process of comparing the result of risk analysis against risk (and often benefit) criteria to determine the significance and acceptability of the risk.

*Risk framing* – the initial assessment of a risk problem, clarifying issues, and defining the scope of subsequent work.

*Risk governance* – the application of governance principles to the identification, assessment, management, and communication of risk. Governance refers to the actions, processes, traditions, and institutions by which authority is exercised and decisions are taken and implemented.

*Risk-informed decision-making* – a purposeful process that uses a set of performance measures, together with other considerations, to 'inform' decision-making.

*Risk management* – activities to handle risk such as prevention, mitigation, adaptation, or sharing. It often includes trade-offs between costs and benefits of risk reduction and choice of a level of tolerable risk.

*Risk metric* – the triplet (C', U', K), where C' is a measure of some consequences C for a specific value V resulting as an outcome from a given activity A (e.g., positive or negative impact), U' represents a measure of uncertainty U associated with C' (e.g., probability or possibility), and K is the background knowledge that supports evaluation of C' and U' (including a judgement of the strength of this knowledge).

*Risk perception* – the subjective judgement people make about the risk, and may vary from person to person.

*Risk source* – a factor or an agent which alone or in combination with other factors (agents) has the potential to give rise to some specified consequences.

*Robustness* – the ability of a system to resist change without adapting its initial stable configuration.

*Safety* – ability to operate, normally or abnormally, without danger of causing human injury or death and without damage to the system's environment.

*Security* – the freedom from those conditions that can cause loss of assets with unacceptable consequences (e.g., ability to prevent an unauthorized access to, or handling of system state).

*Severity* – the magnitude of the damage, harm, or impact.

*State of a system at $t_i$ moment* – the set of relevant properties which the system has at that time. A set of system states may be infinite, but in extreme cases it can be reduced to two states important for the observer (e.g., suitable – unsuitable; moving – resting; meeting the requirements – not meeting the requirements).

*State of a system's environment at $t_i$ moment* – a set of the relevant properties of its environment at that time.

*Static system* (one-state system) – one to which no events occur, therefore it is unchanging over time.

*Survivability* – capability of a system to fulfil its mission, in a timely manner, in the presence of disruptive events.

*System* – a set of interrelated elements. Thus, a system is an entity which is composed of at least two elements and a relation that holds between each of its elements and at least one other element in the set. Each of a system's elements is connected to every other element, directly or indirectly. Furthermore, no subset of elements is unrelated to any other subset.

*System of systems (SoS)* – a set of systems that results when independent systems are integrated into a larger system that delivers unique capabilities.

*System's behaviour* – a sequential chain of changes to the system occurring because of anteceding events such as reactions, responses, and acts. These changes can therefore initiate other events in the system or its environment.

*Tasks* – sequences of activities or operations performed by the same 'actor' on the same object.

*Threat* – a risk source used in relation to reliability and security application (e.g., damage).

*Trustworthiness* – worthy of being trusted to fulfil whatever critical requirements may be needed for a particular component, subsystem, system, network, application, mission, enterprise, or other entity.

*Uncertainty* – a situation of having imperfect knowledge about the true value of a quantity or the future consequences of an activity, scenario, or event.

*Uncertainty metrics* – a probability (e.g., based on the background knowledge) or a possibility function.

*Vulnerability* – the degree to which a system is affected by a risk source or susceptible to damage, harm, or impact.

*Vulnerability metric* – the disruption impact described by two main indicators: the expected loss of performance, and disruption time.

*Vulnerability to a disruptive event* – the degree to which a system is affected by a disruptive event.

# 1

# Introduction

Today's production and service systems are increasingly supported by smart, disruptive technologies based on artificial intelligence, such as: Big Data Analytics, Industrial Internet-of-Things, Digital Twins, as well as virtual and augmented reality. These are systems composed of cyber, physical, and human parts with fuzzy boundaries between individual components of the whole system, generally called Cyber-Physical-Social Systems (C-P-S Systems). Such complex, human-centric, smart systems are characterized by a topology of complex network structures (e.g., network of networks) that should perform specific functions, in an efficient and effective manner, also in an unpredictably changing environment generating various threats and hazards. Thus, the key condition for the successful delivering of products or services is a comprehensive approach to the entire system, with particular emphasis on their complexity, imperfections of knowledge, and broadly understood risk. The concept of such an approach presented in this book is based on the following basic assumptions:

- The high degree of the system's complexity and the emergent nature of its properties – due to multiple interdependency relations between the system's elements and the environment, it is not possible to determine the behaviour based on inherent properties of the constituent elements.
- Multifaceted issues – a comprehensive approach to this kind of systems requires consideration of three basic aspects, namely:
  - The spatial extension of the infrastructure and its surroundings, where the boundary between the system and its surroundings is usually vague or uncertain;

  - ○ Temporary continuity and variability of processes in terms of the life cycle and sustainability;
  - ○ A holistic methodology from a technical, economic, and socio-ethical perspective.

- Imperfection of knowledge available to the decision-maker – knowledge is based in many cases on uncertain, incomplete, and ambiguous data and information; therefore it is imperfect, otherwise decision-makers are guided in practice by the principle of limited rationality and cognition biases.

- Risks associated with the operation of these systems and resulting from their complexity and the unpredictability of changes in the environment are systemic in nature – this makes traditional risk management methods inapplicable to phenomena such as Black Swans, Dragon Kings, and Tipping Points.

Such assumptions are prompted by a thorough analysis of the literature in question, especially synthetic works such as, "The future of risk assessment" (Zio, 2018) and "The science of risk analysis. Foundation and practice" (Aven, 2020). Zio suggests that we are facing revolutionary changes in the perception of risk issues, which he characterizes in the following six points:

(1) The recognition that the knowledge, information, and data available for analysing and characterizing hazards, modelling and computing risk are substantially grown and continue to do so.

(2) The evidence that the modelling capabilities and computational power available have significantly advanced and allow unprecedented analysis with previously infeasible methods.

(3) The concern that the increased complexity of the systems, nowadays more and more made of heterogeneous elements organized in highly interconnected structures, leads to behaviours that are difficult to anticipate or predict, driven by unexpected events and corresponding emerging unknown systems responses.

(4) The realization that to manage risk in a systemic and effective way it is necessary to consider together all phases of the potential accident scenarios that may occur, including prevention, mitigation, emergency crisis management and restoration, and that this entails and extend vision of risk assessment for an integrated framework of business continuity and resilience.

(5) The acknowledgment that risk varies significantly over time and so may also the conditions and effectiveness of the prevention, protection, and mitigation measures installed.

(6) The consideration of the need of solid framework for safety and security assessment of cyber-physical systems.

Aven proposes a comprehensive treatment of all risk issues in the form of a new scientific discipline, 'Risk Analysis' (Aven, 2020). Among other things, he states eight principles on which an effective and efficient approach to risk situations should be based. These are:

(1) The proper risk level is result of a value, context, and evidence/ knowledge-informed process, balancing different concerns.

(2) The process of balancing different concerns can be supported by cost-benefit methods, with broader judgements of risk and uncertainties.

(3) To protect values like lives and heath, as well as environment, the associated risk must be judged to be sufficient low.

(4) Risk perception needs to be incorporated into risk governance, but with great care.

(5) The major strategies are needed for managing or governing risk: risk-informed, cautionary/precautionary, and discursive strategies is also referred to as a strategy of robustness and resilience. In most cases, the appropriate strategy would be a mixture of these three strategies.

(6) Governments should be open and transparent about their understanding of the nature of risks to the public and about the process they are following in handling them.

(7) Governments should seek to allocate responsibility for managing risks to those best placed to control them.

(8) Intervention is needed in the case of market failure or equity issues.

The third pillar on which this work is based is the assumption that in the case of imperfect knowledge which cannot be fully formalized, the lessons learned from studying accidents and disasters that happened in the past (e.g., Moura et al., 2016) are a very valuable source of information. On these foundations, the book was developed, structured into four parts that answer the following main questions:

I. What concept to use to define the research objects of this work?

II. What attributes of these objects are critical to achieving their goals, and what tools are best suited to do so?

III. What methods should be used to study these objects and their behaviour in various situations, especially unusual and risky ones?

IV. How to solve the problems of ensuring the trustworthy operation of these facilities?

Part I answers the first question and is titled: Development of Systems Engineering Concepts: From Single Systems Engineering to Cognitive

Systems Engineering. It begins with a brief overview of basic concepts and ideas of 'System Science' the main principles used in the systems approach as well as the specifics of systems thinking (Chapter 2). It presents a general model of the system, analysing both its structural and behavioural aspects, and proposes a division of systems into four types (passive, reactive, responsive, and active) according to their response to input and their relation to a given goal. The following sections define the model of a system and its main attributes based on Ackoff's idea of a 'system of system concepts' and on this basis proposes a description of cyber-physical systems. It analyses Stafford Beer's concept of animate systems and his Viable System Model as a starting point for defining cognitive social systems, and discusses three basic measures of the system efficiency, namely: actuality, capability, and potentiality, as well as the relationships between these measures. The final section of the chapter defines the basic characteristics of the system performance, namely: performance, productivity, and latency, and points out possible structural, functional as well as information pathologies occurring in the system.

Chapter 3 presents the fundamentals of 'Systems Engineering' as a practical implementation of the systems approach and systems thinking into the field of technical applications. A proposal for a hierarchical structure of concepts in the field of systems engineering is suggested and, on this basis, the author's definition of General Systems Engineering is developed. The main part of the chapter is devoted to the life cycle model of the system engineering process. The model is based on an iterative improvement process with built-in feedback and consists of 13 steps. These steps cover the entire life cycle of the system, from the definition of its basic objectives, through the analysis of feasibility, requirements and functionality, the synthesis and integration of the different parts of the system, to its disposal. A new classification of maintenance policies according to seven main criteria and numerous examples of quantity requirements for selected Technical Performance Measures is proposed. It has been shown that only the holistic and process-oriented approach to Systems Engineering can ensure all essential requirements, namely functional as well as cost and environmental.

Chapter 4 deals with the further development of 'Systems Engineering' into more complex systems, namely System of System Engineering. The basic properties distinguishing complex systems are discussed, especially the concepts of synergy and multidimensional complexity. Different aspects of complexity are analysed, such as structural, spatial, temporal, and disciplinary. On this basis, the concept of System of Systems (SoS) is introduced as a mix of independently operating and actively interacting large systems, integrated with complex goals. A general mathematical model for SoS-type systems and detailed models for network structures, typical of complex network organizations, have been proposed. The following part of the chapter is devoted to Engineered System of Systems, and particularly to the process of creating such systems, i.e., architecting. The purpose of the architecting process is to provide the required properties to the created systems by incorporating into it the so-called metasystem that fulfils a governance role. The functional structure of the metasystem is proposed and the basic functions and sub-functions that a properly designed metasystem should fulfil are characterized.

The purpose of Chapter 5 is to define the main features that distinguish 'Cognitive Systems Engineering' from other systems engineering. The first section is devoted to the discussion of basics of Cognitive Science and Cognitive Systems, and in particular its interdisciplinary nature and relationships between cognitive systems and major classes of systems. This is followed by a brief analysis of cognitive processes of knowledge acquiring in living systems as a chain consisting of basic, meta, and higher cognitive functions. It provides processes of knowledge acquiring in engineered systems which is described and analysed based on a modified knowledge pyramid, and a simplified model of knowledge acquisition process in engineering systems is proposed. The main part of the chapter is dedicated to defining the Cognitive Systems Engineering based on imperfect knowledge acquisition and management concept. The model of the knowledge extraction process as a chain of actions and activities that must be performed to obtain knowledge from raw data is proposed. The chapter concludes by showing the relationship of Cognitive Systems Engineering to other areas of systems engineering and proposing a structural model for its description.

The second question is to be answered by Part II, titled: Concepts of Attribute-oriented Systems Engineering: From Reliability Engineering to Cognitive Dependability Engineering. It begins with Chapter 6, which presents 'Reliability Engineering as the Concept of Failure-free Operation'. This chapter provides an overview of the major concepts that underpinned the emergence and dynamic development of Reliability Engineering as an interdisciplinary field of knowledge. Physical aspects of reliability are also discussed, as being particularly relevant to the engineering approach, and neglected in most works on Reliability Engineering. The mathematical basis for the evaluation of the reliability of elements and entire systems was presented, both based on mathematical statistics and on the theory of probability. Models of selected discrete probability distributions as well as continuous probability distributions were presented, and the calculation formulas necessary to determine the parameters of these distributions are summarized. Selected examples of metrics for reliability attributes have been proposed and illustrated with examples. The section on structural reliability discusses the Reliability Block Diagram method and its application to assess the reliability of systems with different functional structures. The computational models for selected redundant structures were presented, namely so-called $(k, n)$ systems, as well as with hot, warm, and cold reserve. The critical conclusion presents the limitations of the application of these theoretical models in real operating conditions.

Chapter 7 discusses 'Safety Engineering as the Concept of Effective Protection'. It analyses the development of safety issues from the dawn of time to the present day. It presents the most important definitions and assumptions on which the foundations of safety science were built. Three most important trends in the construction of hazard and accident occurrence models were discussed: simple linear accident models, complex linear accident models, and complex non-linear accident models. The advantages and disadvantages of models based on metaphors such as the Domino Effect, Swiss Cheese, and epidemiological model are presented. The second part of the chapter presents the most important Safety Engineering methods and tools. The limitations of using classical models based on the principle of causality in practice are analysed, and modern complex models are presented against this background. It focuses on two different approaches to the safety of complex systems, namely the Systems-Theoretic Accident Mode and Processes/Systems Theoretic Process Analysis (STAMP/STPA) and Functional Resonance Accident Model (FRAM). Their characteristics and basic advantages are presented, as well as the areas of practical applications and development trends are indicated.

'Security Engineering as the Concept of Cyber-security' is the subject of Chapter 8. This chapter addresses the problems of ensuring the security of complex systems, particularly cyber systems. The introductory section discusses the concepts of trustworthy secure systems, which create the foundation for functioning as intended also under the conditions of disruptions, hazards, and threats, with respect to given constraints, limitations, and uncertainty. The following sections presents the idea of the life-cycle-based System Security Engineering, which refers to all processes and activities associated with the system throughout its entire life, with the focus on specific security considerations. Based on these assumptions, the systems security engineering framework was developed, which provides a conceptual view of the key contexts of the systems security engineering activities, both technical and nontechnical. The next section focuses on cyber security as a process for protecting information by preventing, detecting, and responding to attacks. The cyber security concept called CIA (Confidentiality-Integrity-Availability) triad is discussed, and in conclusion, a general model for securing resources from adversarial attacks through appropriate security policies and selection of appropriate countermeasures is presented.

Chapter 9 is devoted to 'Resilience Engineering as the Concept of Process Continuity'. The topic of the chapter is the concept of resilience and its application in science and practice, especially in engineering. The foundations of a new approach to creating, implementing, and operating systems based on the so-called resilience thinking and its characteristics, including adaptability and capacity for transformation, are presented. A brief review of main development trends of the resilience approach in various areas of application is made and common features of these studies as fundamental for the emergence of a new discipline of knowledge – the Resilience Science – are summarized. First part of the chapter concludes with a proposal for a matrix of strategies and aspects for providing resilience to systems and organizations. Next part of the chapter is devoted to a discussion of the principles, methods, and models used in Resilience Engineering. The basic relationships on which modelling the resilience of systems and organizations are based and are presented, and five main types of models used in qualitative and quantitative research on resilience are analysed. At the end of the chapter, a universal simplified functional model of resilience is proposed, which will be a starting point for the development of complex quantitative-qualitative models in the following chapters of this work.

'Dependability Engineering as the Concept of Fault-tolerant Functioning' is the subject of the Chapter 10. The chapter provides a concept of Dependability Engineering as a relatively new, still dynamically developing field of knowledge. A historical outline of the development of this concept is presented, starting with the idea of failure tolerance functioning and fault-tolerant systems. Based on these ideas, the Dependable Computing concept has matured and has been successfully applied to the field of computer science and information and communication technology. The second part of the chapter presents the basics of Dependability Engineering and the five selected approaches for building the foundation of this novel, emerging applied science discipline. Each of these approaches have their advantages and disadvantages, so their use should depend on the application area and the specifics of a given type of system. The vulnerability-based concept as well as the risk-related concept seems to be the most universal and therefore, in the remainder of this work an attempt will be made to develop a universal model of Cognitive Dependability Engineering for cyber-physical-social systems on these assumptions.

The final chapter of Part II is Chapter 11 about 'Cognitive Dependability Engineering as the Concept of Trustworthy Performance'.

This Chapter is devoted to an extended version of Dependability Engineering, namely Cognitive Dependability Engineering (CDE). It deals with systems of very high complexity, called human-centric cyber-physical-social systems, and the research perspective of these systems is focused on the continuity of operation and functioning. At the core of the CDE is the concept of trustworthy performance, and the research method recommended for such defined cognitive objects is transdisciplinary. The chapter firstly discusses key features of scientific theories, and conditions of scientific cognition according to the Einstein's model. Then the process of emergence of a new science was suggested, which can be divided into three main stages. According to this model, it has been proposed the general concept of Cognitive Dependability Engineering, and its area of applicability depending on the complexity of the systems and the level of uncertainty. The foundations of the elicitation procedure as a basis for cognitive dependability assessment of complex systems are presented, and a pre-paradigm for a new field of knowledge, Cognitive Dependability Engineering, is proposed.

Part III is titled: Modelling and Simulation the Operation of Cyber-Physical-Social Systems in a Risky Environment. Its intention is to answer the third question. Chapter 12 provides the fundamentals of Modelling and Simulation of Complex Engineered Systems'. The model is understood as a surrogate for the real system and represents its characteristics and attributes in experimental studies. The modelling process is a constructive activity whose goal is to build a sufficiently good model. The chapter presents the algorithm of the entire modelling and simulation process and discusses all its steps. Then the process of creating mathematical models divided into seven steps is demonstrated. The basic advantages and disadvantages of the modelling and simulation process are analysed. The second part of the chapter is devoted to development and application of the modelling and simulation process. The life cycle of a modelling and simulation process was presented and examined. The relationships between the conceptual model, the working model, and the real complex engineered system is discussed. The eight key factors contained in the examination of results credibility are demonstrated and scrutinized. The chapter concludes with an analysis of the risks associated with uncertainty regarding the credibility of modelling and simulation results, as well as criteria for deciding whether to accept or reject these results.

Chapter 13 discusses 'Modelling of Cyber-Physical-Social Systems'. This chapter presents the basics of Cyber-Physical-Social Systems modelling. To describe the structure of these systems, methods based on multi-agent models have been proposed, with an autonomous agent as the basic element. The main advantages of multi-agent modelling are described, as well as IT tools to support the modelling processes.

The following sections discuss how to model the processes carried out within the Cyber-Physical-Social Systems. The most widely used process modelling standards are presented, namely: Event-driven Process Chain model, Entity Relationship Diagram, Data Flow Diagram, Petri networks, and Value Stream Mapping. The third part of the chapter presents methods for modelling the risks of loss of process continuity within complex systems of the Cyber-Physical-Social type. A new classification of risk sources was proposed from the following perspectives: human resources, financial resources, intangible assets, and infrastructure resources. In the final section of the chapter, a model based on a power distribution is presented as particularly suitable for describing extreme phenomena occurring, for example, in natural disasters.

'Simulation of Cyber-Physical-Social Systems behaviour in a Risky Environment' is the topic of Chapter 14. The main purpose of modelling systems and processes, especially complex ones of the Cyber-Physical-Social type, is to create opportunities for simulation studies. It presents the basics of the methodology for building simulation models and using them to generate various cases that have not yet been observed in practice but may occur in the future. This particularly applies to hazardous events that pose a threat to the environment, human safety and the continuity of production or service processes. The first section discusses the principles of the two basic methods on which simulation models are based, namely System Dynamics Modelling and Simulation (SDMS) and Discrete Event Simulation (DES). The second section is devoted to a practical example of simulating a complex system behaviour under uncertainty. The main objective of this example is to demonstrate the feasibility of using the proposed framework to assess the vulnerability and resilience of global supply networks. The subject of the research was a commodity steel plant located in Central Europe, and the main goal of the research was to determine the recommended actions of decision-makers (modelled using a multi-agent technique) in the occurrence of disruptions in the supply of raw materials necessary for steel production.

The last question is to be answered by Part IV, entitled 'Managing Risks in Cyber-Physical-Social Systems under Deep Uncertainty'. It begins with Chapter 15 – 'Uncertainty-oriented Concepts of Decision-making'. Decision-making under conditions of uncertainty is an important and responsible process, especially if it concerns the continuity of operation of complex Cyber-Physical-Social Systems type structures. The chapter presents the concept of risk because of the imperfect knowledge available to the decision-maker under conditions of uncertainty. Decisions can be made about either taking or not taking given actions or activities that will take place in an uncertain environment. The value of risk depends on the strength of knowledge about the object of the decision available to the decision-maker, the uncertainty associated with the possibility of a given outcome, and the value of consequences of a given outcome. The second part of the chapter deals with the concept of risk-informed decision-making (RIDM) as a purposeful process that uses a set of performance measures, together with other considerations, to inform the decision-maker about the risks associated with the consequences of the various decision options. The third section of the chapter proposes the concept of dependably related decision-making in the form of a 10-step algorithm. This concept forms the basis of a generalized cognitive dependability model and Cognitive Dependability Governance Framework for Cyber-Physical-Social Systems operating under deep uncertainty presented in next chapter.

Chapter 16 deals with 'Cognitive Dependability-based problem-solving in Cyber-Physical-Social Systems Operating under Deep Uncertainty'. Decision-making under conditions of uncertainty is an important and responsible activity, being the final step in the problem-solving process. In real-world practice, the entire process is fraught with uncertainty, with the degree of uncertainty having a significant impact on the risk involved in making the final decision. In the first part of the chapter, general concepts and principles related to the consideration of uncertainty in the problem-solving process are presented with particular emphasis on the decision-making stage. The second part is devoted to the same issues related to conditions of deep uncertainty. It has been proposed to use the concept of a Digital Twin to support the process of problem solving under deep uncertainty using an interactive dialogue system between a human expert, representing natural intelligence, and a Cognitive Digital Twin, representing artificial intelligence. The last section of the chapter presents a framework for cognitive dependability based problem-solving in Cyber-Physical-Social Systems operating under deep uncertainty. The starting point for the development of this framework is a simplified model of a Cyber-Physical-Social System cognitive dependability, and a general model for solving operational problems under deep uncertainty based on the concept of cognitive dependability, which are presented in both graphic and descriptive form.

Chapter 17 shows 'Application Examples as Lesson Learned from the Case Studies'. The chapter presents examples of three different disasters that have occurred over the past several decades, and documentation on the analysis of their course and potential causes is widely available. The first example concerns the collapse of a bridge on 1 August 2007, a stationary system that is part of the transportation infrastructure. In this case, the cause of the catastrophe was not unpredictable forces of nature, but the lack of a comprehensive analysis of the entire complex system and all processes taking place in it, both in terms of transport, repair, and logistics. The second example applies to the Three Mile Island nuclear accident on 28 March 1979. This catastrophe inspired Charls Perrow to develop the Theory of Normal Accidents, one of the conclusions of which is that human cognitive limits are important in causing system accidents. In the third example, an analysis was performed of the damage from the Great East Japan Earthquake and Tsunami on 11 March 2011. This example shows that classical probabilistic methods used in the assessment of extreme events of the natural disaster type cannot be the basis for decision-making in the case of high-impact hazards. However, in these situations, methods based on the precautionary and resilient principles should be used.

The last Chapter 18 is to summarize the whole work and to propose 'Conclusions and Concluding Considerations'. The work concludes with an Index of key terms and a glossary of definitions of concepts used, the meaning of which may be unclear or ambiguous.

## References

Aven, T. (2020). *The Science of Risk Analysis. Foundation and Practice*. London and New York: Routledge, Taylor and Francis Group.

Moura, R., Beer, M., Patelli, E., Lewis, J. and Knoll, F. (2016). Learning from major accidents to improve system design. *Safety Science*, 84: 37–45.

Zio, E. (2018). The future of risk assessment. *Reliability Engineering and System Safety*, 177: 176–190.

# Part I

# Development of Systems Engineering Concepts

## From Single Systems Engineering to Cognitive Systems Engineering

# 2
# Basics of Systems Science[*]

The chapter introduces the remaining chapters of Part I by defining and analysing the basic concepts and ideas of Systems Science. It discusses the main principles used in the systems approach and the specifics of systems thinking. It presents a general model of the system, examining both its structural and behavioural aspects and proposes a division of systems into four types (passive, reactive, responsive, and active) according to their response to input and their relation to a given goal. The following section defines the model of a system and its main attributes based on Ackoff's 'system of system concepts' and on this basis proposes a description of cyber-physical systems. It analyses Stafford Beer's concept of animate systems and his Viable System Model as a starting point for defining cognitive social systems, and discusses three basic measures of the system efficiency, namely, actuality, capability, and potentiality, as well as the relationships between these measures. The final section of the chapter defines the basic characteristics of the system performance, namely, performance, productivity, and latency, and points out possible structural, functional and information pathologies occurring in the system.

## 2.1 Main Principles of Systems Science – The Systems Approach

As the complexity of natural and artificial objects increased, scientific methods based on the General Principle of Causality, deductive reasoning and reductionism became insufficient to describe and explain the properties and behaviour of complex structures, consisting of elements linked by mutual relations. Breakthroughs leading to the emergence of new paradigms of science to circumvent these limitations were the laws

[*] https://orcid.org/0000-0002-2630-3507.

of thermodynamics with the concept of entropy, Max Planck's quantum theory, Werner Heisenberg's principle of ambiguity, and Niels Bohr's principle of complementarity. They contributed to changing the attitude to *uncertainty* from negative, identifying uncertainty with ignorance, to positive, considering it an important part of knowledge about the world around us.

Also, in the field of biological and social sciences, in the first half of the 20th century, phenomena that could not be explained by classical scientific methods were discovered. The main representative of the new trend in biological sciences was Ludwik von Bertalanffy, considered to be the creator of *the Systems Theory*, who presented his concept in universal terms for the first time in the journal *Science* in 1950, writing among others that the properties and modes of operation at higher levels of the organization cannot be explained by summing up the properties and modes of operation of their components examined separately. However, when we know the set of components and the relationships between them, higher levels of organization can be explained by their components (von Bertalanffy, 1950).

The third pillar on which the new scientific paradigm, called systemic, was built is cybernetics. This notion was introduced in 1948 by Norbert Wiener (Skyttner, 2008) and since then cybernetics has become a buckle that binds together the science of the animated and inanimate world. The analogies between these worlds were based on Control Theory (CT) and Communication and Information Theory (CIT), and the computer became the primary experimental tool.

In this situation, in the second half of the twentieth century a new meta-discipline of science emerged, called *Systems Science*, based mainly on holism, synthesis, reduction and inductive inference. Systems Science is characterized by a specific point of view of the world, consisting in the pursuit of understanding man and his environment as mutually interacting elements of one system, considered from many perspectives and in various aspects, but always in a holistic perspective. Based on the Systems Science, various system theories have developed, including the General Systems Theory, which nowadays plays the role of a universal language that brings together different areas of interdisciplinary communication. General Systems Theory is still in a phase of continuous development, aiming at laying the foundations of universal science, based on the so-called *Law of Rights*, which will integrate all areas of scientific knowledge. Its role was aptly formulated by Keneth Boulding (Boulding, 1956) in the following way:

**General Systems Theory is the skeleton of science in the sense that it aims to provide a framework or structure of systems on which to hang**

**the flesh and blood of particular disciplines and particular subject matters in an orderly and coherent corpus of knowledge.**

Systems science have developed a new scientific paradigm called holism, that is, *systems approach*. This approach looks at the whole through the role and function of parts in the whole, considering the relationship of cause and effect (often undisclosed, non-linear, and distant in time). The systems approach is characterized by the following important shifts (Bukowski, 2019):

- from an emphasis on the part to focusing attention overall,
- from a static structure to a dynamic and process structure,
- from a metaphor of a hierarchical structure or a linear chain to the metaphor of a network,
- from the desire for an absolute accuracy and precision to the approximate description.

The basis for the systemic approach is the so-called *systems thinking* (Gharajedaghi, 2006; Klir, 1969 and 1991; von Bertalanffy, 1962, Weinberg, 1975), based generally on synthetic research procedure which can be reduced to three following main steps (Bukowski, 2019):

- identification of the entire system, which includes the examined elements (e.g., subsystems),
- evaluation and description of the system's properties and its behaviour in time,
- determination of the properties of an individual system's elements and their behaviour in time, considering the impact of these elements on the functioning of the whole system.

The essential aim of this type of synthesis is not a precise identification of the system's structure, but rather gaining knowledge about its behaviour in time and particularly about its functions. Thus, the *cognitive emphasis* of systemic synthesis is placed at explaining and understanding the behaviours and functions of the system, with a specific emphasis on time variation, while an analytical approach focuses on fixed in time structural characteristics of the system.

The systems approach and thinking became the starting point for formulating the general principles that characterize Systems Science. The systems approach and thinking became the starting point for formulating the general principles that characterize Systems Science. The most important of these can be reduced to the following attributes of systems:

a. *Interdependence of elements*. Elements, even very numerous, but not interconnected and independent from each other, cannot create a system.

b. *Holism.* The property of the system cannot be determined based on the the properties of its elements without testing the system overall.

c. *Purposefulness.* Interactions between the elements of the system lead to the achievement by the system of a specific goal, which may be a given final state, or a certain equilibrium state.

d. *Transformation process.* In all systems, inputs are transformed into outputs. In natural systems, this transformation is cyclical.

e. *Closed and open systems.* In closed systems, the inputs are unambiguously fixed, while in open systems, there are additional inputs from the environment, depending on its possible changes.

f. *Entropy and degradation.* Artificial systems strive to maximize their entropy over time, which corresponds to an increase in the level of indeterminacy (disorder and randomness) and system degradation. Natural (animated) systems import energy from the environment, which allows to stop degradation of the system in a certain, limited period, thanks to the creation of the so-called negentropy.

g. *Regulation and control.* The components that make up the system must be controlled using feedback to achieve the intended state. In the case of live systems, this is a dynamic equilibrium state.

h. *Hierarchy.* Complex systems may have a hierarchical structure, which means that individual systems may be components of other systems.

i. *Differentiation and specialization.* In complex systems, their elements may perform special functions or execute parts of processes (tasks, procedures).

j. *Equifinality and multifinality.* Open systems may have convergent properties, i.e., achieving the same goals under different initial conditions, or divergent properties, if they may reach different end states from a specific initial state.

## 2.2 Defining System and its Attributes – The Concept of Cyber-physical Systems

Based on these attributes, many attempts have been made to define the term 'system'. George J. Klir made a critical analysis of the applied ways of defining systems and proposed a new methodology, which he named *inductive approach,* as opposed to deductive approach dominating in that time. This approach was based on four basic features of systems, namely: the resolution level of examined system values (i.e., the accuracy and frequency of measurement), behaviour over time (variability of system values observed over time), relations stable over time (the relationships between system values that are stable over time) and properties describing these relations. This resulted in a division into the following five classes of definitions (Klir, 1969):

I. A given system is a set of values examined at a certain resolution level.

II. A system is a set of values representing the variability of values under consideration over time.

III. A system is a given relation, stable over time, between instantaneous and/or past and/or present external values.

IV. A system is a given set of elements and their regular behaviour and the set of forces between individual elements and between the elements and the environment.

V. A system is a set of states and a set of transitions between states, where these transitions may be indeterministic in nature.

From the point of view of practical applications, the last two definitions have the best application properties – the first one (no. IV) is called a UC definition (Universe of Discourse and Couplings), whereas definition no. V is an ST definition (State Transition). Both definitions are similar in nature and can be described in one general *structural model* in the following form:

$$S = (E \times R) \tag{2.1}$$

where:  $S$ – modelled system,

$E$ – elements of the system,

$\times$ – Cartesian product,

$R$ – relations between system elements.

This model can be regarded as universal, if we assume a generalized interpretation of the symbols $E$ and $R$. Then the $E$ elements of the system can be its parts (material or intangible objects), events, states, or functions, whereas $R$ relations may apply both to relationships between different elements (e.g., in hierarchical systems), and forces such as mutual interactions and interactions in dynamic systems (e.g., in a process approach – flows of matter, energy, and information).

In 1971, Russell L. Ackoff proposed to organize the systems concepts by publishing an article titled "Towards a System of Systems Concepts" in the *Management Science Journal* (Ackoff, 1971). The publication is a framework, which comprises 32 points defining the basic concepts in the theory of systems, the relations between them, and a classification of systems in a behavioural approach. The concept of system has been defined as follows:

**System is a set of interrelated elements. Thus, a system is an entity which is composed of at least two elements and a relation that holds between each of its elements and at least one other element in the set.**

**Each of a system's elements is connected to every other element, directly or indirectly. Furthermore, no subset of elements is unrelated to any other subset.**

The basic division of systems, based on their relation to reality, distinguishes two types of systems: abstract and concrete. *Abstract systems* are characterized by the fact that all items included in their composition are intangible. Relations between abstract elements are defined using assumptions in the form of postulates or axioms. Examples of such systems include natural and programming languages, and philosophical and classification systems. *Real systems* are those in which at least two elements are material objects. They are a study subject for *Systems Engineering* (SE), which is a field of application of knowledge based on systems theory and the subject of interest of this paper.

Every concrete system, at any moment in time, can be described by a set of its properties (characteristics). *The state of a system* at $t_i$ moment is the set of relevant properties which the system has at that time. A set of system states may be infinite, but in extreme cases it can be reduced to two states important for the observer (e.g., suitable – unsuitable, moving – resting, meeting the requirements – not meeting the requirements). *The environment of a system* is a set of elements and their relevant properties, which are not part of the system (they are outside its borders) but may influence (substantially) the state of the system. *The state of a system's environment* at $t_i$ moment is a set of the relevant properties of its environment at that time. Depending on the severity of the environment's impact on a given system, systems are divided into closed and open systems. A *closed system* is one that has no environment, that is, no element outside of the system has a significant impact on the state of the system. The state of an *open system* depends on both the properties of the system and its environment.

The properties of the system and its environment and more specifically their values may change over time. Such important changes to at least one structural property of the system and its environment are called *events*. A *static system* (one-state system) is one where no events occur, therefore it is unchanging over time. *Dynamic systems* (multi-state systems) are ones where events occur and whose state changes over time. The events which have a decisive influence on the changes taking place in systems are reaction, response, and act. A *reaction* of a system is an event for which another event, preceding it, that occurs to the same system, or its environment is necessary and sufficient. A *response* of a system is an event for which another event, preceding it, that occurs to the same system or to its environment is necessary but not-sufficient. For a system response to take place it is therefore necessary that an additional stimulating event, a co-producer must occur in another system or its environment. *An act* of a system is an event that occurs without external stimuli as a self-determined action caused by the system itself.

*A system's behaviour* is a sequential chain of changes to the system occurring because of preceding events such as reactions, responses, and acts. These changes can therefore initiate other events in the system or its environment. To describe the behaviour of a system it is necessary to know both the significant events preceding a given moment, as well as the consequences of these events. Depending on how the system reacts to these events, or behaves, four basic kinds of systems can be distinguished: *passive* (with a static behaviour), as well as *reactive, responsive,* and *active* (with a dynamic behaviour) systems. Table 2.1 contains a short description of behaviour and relation to goal of these individual types of systems, with simple examples. From the standpoint of practical application in production and services, including logistics, the *goal-seeking* and *purposeful systems* are particularly interesting. Smart and creative control and management of such systems require the use of structures with autonomous elements, connected by relations allowing for mutual communication in real time.

**Table 2.1:** The basic types of systems and their characteristics (based on Bukowski, 2019).

| Type | Behaviour | Relation to Goal | Example |
|------|-----------|------------------|---------|
| Passive | static | fixed | tool |
| Reactive | state-maintaining | variable (given) | thermostat |
| Responsive | goal-seeking | variable (chosen) | autopilot |
| Active | purposeful | adaptive | smart grids |

*The purposefulness* (or in the case of operational systems – functioning) of these systems is based on pursuing a certain state which can be defined in a strategic and operational scale. The notion of *functioning* is understood as a manifestation of rational behaviour, involving the fulfilment of certain (required) functions, e.g., the execution of production processes or services. The purpose of an operational system should be to fulfil the expectations of the object of the action (owner, customer, society). *The goal* of an active system is the preferred state of a system, guaranteeing the production of a given functions at the outcome within a specified, relatively short time-period. *The objective* of an active system is the preferred state of the system, guaranteeing the production of a given function in an ultimately desirable degree in a long period of time (a long-term goal).

*The function of a system* is to generate (produce, achieve) results in line with the goals and objectives of the system, regardless of the state of the system and its environment. In practice, the function of production systems is the effective and efficient manufacturing of certain products, whereas service systems come down to providing services in accordance with the requirements of the customers at prices (costs) favourable to both sides. The classification of systems according to the basic criteria such as the origin of the system, its relation to reality, its content, relation to the environment, degree of determinacy, complexity, variability in time, continuity and role are presented in Table 2.2.

**Table 2.2:** The classification of systems according to the basic criteria (Bukowski, 2019).

| Criterion | Dichotomous Division | |
|---|---|---|
| Origin | natural | artificial |
| Relation to reality | real | abstract |
| Substance | material | immaterial |
| Relationship with the environment | closed | open |
| Degree of determinacy | deterministic | indeterministic |
| Degree of complexity | single | multiple |
| Variability in time | static | dynamic |
| Continuity | discrete | continuous |
| Fulfilled function | operating | configuration |

According to the above classification, a general division of the systems into the following six classes is proposed: living systems, natural physical systems, artificial physical systems, sociological systems, abstract systems, and cyber systems. In practice, a simplified division into physical systems (natural and artificial), cyber systems (ICT and abstract), and social systems (living and sociological) is often used, and systems of the highest complexity, which are a conglomerate of all these systems, are called *Cyber-Physical-Social Systems*. The following is a brief description of the proposed structure for subdividing the systems using the criteria summarized in Table 2.2:

- Cyber Systems
  - Abstract Systems – artificial, abstract, immaterial, configuration,
  - ICT Systems – artificial, real, material, immaterial, discrete, operating.
- Physical Systems
  - Natural Physical Systems – natural, real, material, open, indeterministic, multiple, dynamic,
  - Artificial Physical Systems (engineered) – artificial, real, material, open or closed, deterministic or indeterministic, static or dynamic, discrete, continuous, operating.

- Social Systems
  - Living Systems – natural, real, material, open, indeterministic, multiple, dynamic, continuous, operating,
  - Sociological Systems – natural, real, material, immaterial, open, closed, indeterministic, multiple, dynamic, continuous, operating.

In business practice, *organizations* comprise a very important class of systems. It is proposed the following general definition of an organization: "An organization is a controllable, function-producing and goal-seeking entirety, which is separate from its environment and whose elements are connected by mutual relations, both physical and intangible" (Bukowski, 2019). The function of an organization is to complete the pursued goals effectively and efficiently. Material relations are the flows of goods (raw materials, semi-finished products, finished products and energy) between the elements of the organization. Intangible relationships are communication, ensuring the flow and exchange of information between the elements of the organization.

Figure 2.1 shows a general model of a system with a defined structure and dynamic behaviour. Its inputs are the events occurring in its environment (controllable – resources supply, as well as uncontrollable – disturbances), and its outputs are the produced functions (controllable – goods and services delivery, as well as uncontrollable – disruptive events). Ackoff's concept has become the basis for defining and investigating *physical systems* and has also found application for *cyber-physical systems*. However, it has not found wider application among researchers working on *living systems* (human and social).

ENVIRONMENT OF THE SYSTEM

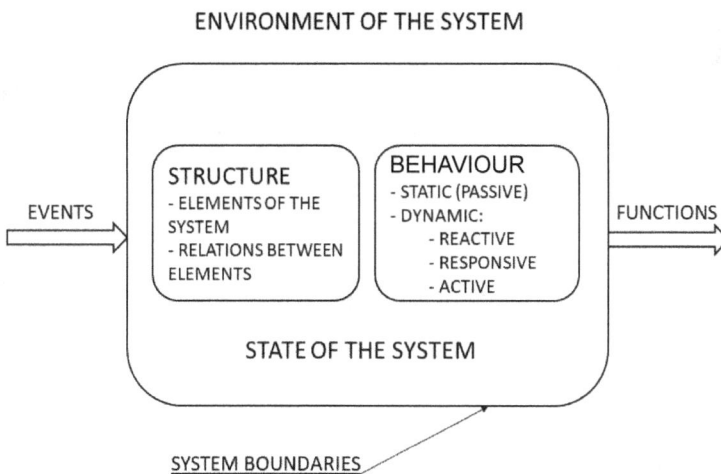

**Figure 2.1:** General model of cyber-physical system according to the concept proposed by Ackoff (1971).

## 2.3  Concept of Living Systems – The Viable System Model

The idea of building a model of an animated system based on the human body was first presented by Stafford Beer in his book *Brain of the Firm* (1972). This concept, called *neurocybernetics*, was developed by Beer in his later works (Beer, 1981, 1985, 1989) and is currently known as *The Viable System Model* (VSM). By 'viability' the author meant the ability of a system to maintain a separate existence and therefore to survive regardless of changes and threats in its environment. The Viable System was mainly characterized by the following features:

- capacity for self-regulation and self-correction,
- ability to learn,
- adaptability, and
- the capacity to develop and evolve.

Fundamental to this concept is the concept of variability and the so-called Law of Requisite Variety which states that the variety of the control unit must be at least the same as the variety of the governed system. A significant reduction in variability is only possible through system recursion, which means that each level of the system (subsystems, assemblies, parts, and elements) is a recursion of its metasystem. Based on these basic assumptions, Beer formulated four general conditions that any viable system must meet:

- variety, diffusing through an institutional system, tends to equate; should be designed to minimize cost and the risk of damage,
- every channel providing information flow between the operational unit, the management unit, and the environment must have a higher capacity than each of these subsystems,
- whenever an information channel crosses a subsystem boundary it shall be processed (e.g., undergoes transduction), the transducer variation shall be at least equivalent to the variation of this channel,
- all three conditions must be maintained during operation without interruption or significant delay.

The conditions mentioned above form the so-called *Recursive System Theorem*, which Beer formulated as follows: "any viable system contains, and is contained in, a viable system".

Based on the above considerations, the so-called Viable System Model (Beer, 1981, 1985) was proposed, which defines the necessary and sufficient conditions for a given system to be considered a viable system. This model consists of five basic systems embedded in the environment, whose interdependencies are shown symbolically in Figure 2.2. System

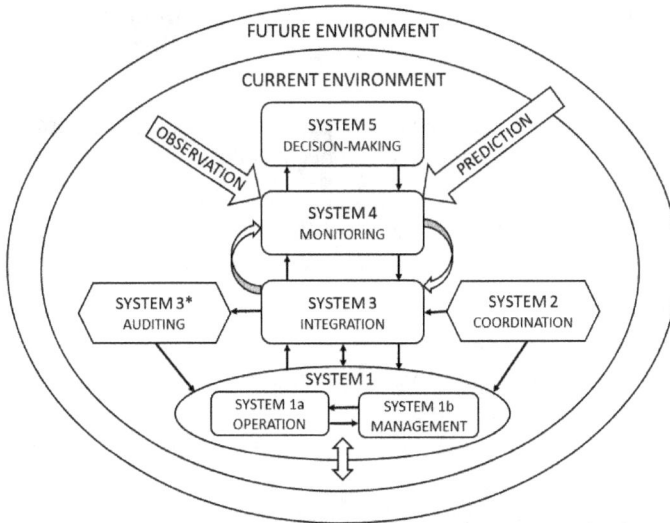

**Figure 2.2:** General model of living systems according to the concept proposed by Beer (1981).

1 is the basic element of the whole, performs production or service tasks and consists of two parts: 1a, the operation subsystem, and 1b, the management subsystem. The system is in constant interaction with the current environment and the 2, 3 and 3* systems. System 2 is responsible for coordinating all the subsystems included in System 1 to ensure that it functions harmoniously. System 3 is the central part of the system and fulfils the role of integrator of systems 1, 2, and 3* activities. System 3* has a control role and is responsible for auditing the results of System 1. The role of System 4 is to continuously supervise and monitor the current environment through observation and measurement and the future environment through forecasting and predicting its probable changes. This system is connected in a closed loop with System 3 and provides the necessary information to System 5, which supports decision-making overall. In organizational systems, the company's policy is set within this system, along with its mission, vision, and development.

The VSM design and implementation process can be presented in a structured form as a sequence of four basic stages (based on Pérez Ríos, 2010):

Step I – recognition of the general structure, specificity, and goals of the system.

Step II – identification of the vertical (hierarchical) structure of the system and its relationship with the complex environment.

Step III – detailed analysis of the vertical structure of the system, with particular emphasis on the elements significantly affecting the viability of the system.

Step IV – checking the degree of coupling of all system's elements at all recursion levels, to ensure full coherence of all parts of the system and its goals.

To verify the results achieved by the system, Beer (1979) proposed three basic measures of the system efficiency, which he defined as follows:

a. Actuality (A) – the current achievement using existing resources and constraints.

b. Capability (C) – the possible achievement with existing resources and within existing constraints.

c. Potentiality (P) – what could be achieved by developing resources and removing constraints.

Relationships between these measures can be used to determine the basic characteristics of the system performance, namely:

$$Performance = \frac{A}{P} \tag{2.2}$$

$$Productivity = \frac{A}{C} \tag{2.3}$$

$$Latency = \frac{C}{P} \tag{2.4}$$

Unsatisfactory values of these measures may indicate pathologies occurring in the system. Pérez Ríos (2012) distinguishes three basic groups of such pathologies, namely:

- Structural pathologies – resulting from improper system topology, e.g., excessively sensitive to changes in the environment.
- Functional pathologies – caused by disturbances in the implementation of the basic functions of the system.
- Information pathologies – related to disruptions in communication and information flow.

The concept of the Viable System seems to be the most appropriate to be used to create, model, and optimize cognitive social systems and therefore it was developed later in the work (see Chapter 5).

## 2.4 References

Ackoff, R.L. (1971). Towards a system of systems concepts. *Management Science*, 17(11): 661–671.

Beer, S. (1972). *Brain of the Firm*. Allen Lane: The Penguin Press.

Beer, S. (1979). *The Heart of Enterprise*. Chichester: Wiley.

Beer, S. (1981). *Brain of the Firm: The Managerial Cybernetics of Organization*. New York: Wiley.

Beer, S. (1985). *Diagnosing the System for Organizations*. Chichester: Wiley.

Beer, S. (1989). The viable system model: Its provenance, development, methodology, and pathology. *In*: R. Espejo and R. Harnden (eds.). *The Viable System Model, Interpretations and Applications of Stafford Beer's VSM*. Chichester: Wiley.

Boulding, K. (1956). General system theory: The skeleton of science. *Management Science*, 1956/2.

Bukowski, L. (2019). *Reliable, Secure and Resilient Logistics Networks. Delivering Products in a Risky Environment*, Springer Nature Switzerland AG 2019. ISBN 978-3-030-00849-9 (Hardcover), ISBN 978-3-030-00850-5 (eBook).

Gharajedaghi, J. (2006). *System Thinking. Managing Chaos and Complexity*. Elsevier.

Klir, G.J. (1969). *An Approach to General Systems Theory*. NY: Van Nostrand Reinhold Co.

Klir, G.J. (1991). *Facets of Systems Science*. NY: Plenum.

Pérez Ríos, J. (2010). Models of organizational cybernetics for diagnosis and design. *Kybernetes*, 39(9/10): 1529–1550.

Pérez Ríos, J. (2012). *Design and Diagnosis for Sustainable Organizations. The Viable System Method*. Berlin, Heidelberg: Springer.

Skyttner, L. (2008). *General Systems Theory: Problems, Perspectives, Practice*. Singapore: Word Scientific.

von Bertalanffy, L. (1950). An outline of general system theory. *The British Journal for the Philosophy of Science*, 1(2)(Aug. 1950): 134–165. http://www.isnature.org/Events/2009/Summer/r/Bertalanffy1950-GST_Outline_SELECT.pdf.

von Bertalanffy, L. (1962). General system theory: A critical review. *General Systems*, 7: 1–20.

Weinberg, G. (1975). *An Introduction to General Systems Thinking*. NY: Wiley.

# 3

# Systems Engineering[*]

This chapter presents the fundamentals of Systems Engineering as a practical implementation of the systems approach and systems thinking into the field of technical applications. A proposal for a hierarchical structure of concepts in the field of Systems Engineering is suggested and, on this basis the author's definition of General Systems Engineering is developed. The main part of the chapter is devoted to the life cycle model of the system engineering process. The model is based on an iterative improvement process with built-in feedback; it consists of 13 steps. These steps cover the entire lifecycle of the system, from the definition of its basic objectives, through the analysis of feasibility, requirements and functionality, the synthesis and integration of the different parts of the system, to its disposal. In the section a new classification of maintenance policies according to seven main criteria and numerous examples of quantity requirements for selected Technical Performance Measures is proposed. It has been shown that only the holistic and process-oriented approach to Systems Engineering can ensure all essential requirements, namely, functional as well as cost and environmental.

## 3.1 Fundamentals of System Engineering

The paradigms of system approach and system thinking developed within the framework of Systems Science have become the basis for the development of *Systems Engineering* (SE), which is the implementation of the holistic principle into practical applications in the field of widely understood technology. Since there is yet no consensus on the adoption of a single, universal definition for the term Systems Engineering, the following will be the proposals put forward by the dominant institutions in this area of applied science.

[*] https://orcid.org/0000-0002-2630-3507.

The International Council on Systems Engineering (INCOSE, 2007) defines it as follows:

**Systems engineering is an interdisciplinary approach and means to enable the realization of successful systems. It focuses on defining customer needs and required functionality early in the development cycle, documenting requirements, and then proceeding with design synthesis and system validation while considering the complete problem. Systems engineering considers both the business and technical needs of all customers with the goal of providing a quality product that meets the user needs.**

In contrast, the proposal made by the Department of Defense (DOD, 2002) is as follows:

**An approach to translate approved operational needs and requirements into operationally suitable blocks of systems. The approach shall consist of a top-down, iterative process of requirements analysis, functional analysis and allocation, design synthesis and verification, and system analysis and control. Systems engineering shall permeate design, manufacturing, test and evaluation, and support of the product. Systems engineering principles shall influence the balance between performance, risk, cost, and schedule.**

The Defense Acquisition Guidebook in the Chapter 3 provides a definition of Systems Engineering as (DAG, 2013):

**Systems engineering (SE) is a methodical and disciplined approach for the specification, design, development, realization, technical management, operations, and retirement of a system. ... A system is an aggregation of system elements and enabling system elements to achieve a given purpose or provide a needed capability. The enabling system elements provide the means for delivering a capability into service, keeping it in service or ending its service, and may include those processes or products necessary for developing, producing, testing, deploying, and sustaining the system.**

Although the above three definitions differ from each other, they share a high degree of similarity, which can be reduced to the following characteristics:

- Systems Engineering is a universal *methodological approach* with a process character,
- it is *interdisciplinary*, requiring an understanding of different areas of knowledge and the relationships between them,
- its main task is the *transformation* of operational requirements (customer needs) into an integrated system solution design,

- this process follows the *top-down* rule in an iterative way through all phases of the project to balance the basic properties of the system, i.e., performance, quality, costs, risk, time (schedule), and human factors (usability, safety, etc.), and

- the predominant orientation is the *life cycle model*, which consists in considering all relevant phases of the life of the designed system, starting from the identification of needs until the disposal of the worn-out system.

Based on the above, it is proposed to formulate a generalized definition of Systems Engineering in the following form:

GENERAL SYSTEMS ENGINEERING IS AN INTERDISCIPLINARY, METHODICAL APPROACH WITH AN ITERATIVE NATURE AND A LIFE CYCLE ORIENTATION THAT ENABLES THE EFFICIENT AND EFFECTIVE SOLUTION OF COMPLEX DESIGN PROBLEMS UNDER CONDITIONS OF UNCERTAINTY AND RISK.

This definition is universal, emphasizing the interdisciplinary nature of Systems Engineering, its holistic life-cycle-oriented iterative approach, and the need to solve complex problems efficiently and effectively under conditions of uncertainty and risk.

Figure 3.1 shows a proposal for a hierarchical structure of this concept in the field of Systems Science which is divided into theoretical and applied sciences. In the theoretical sciences, there are several versions of General Systems Theory (e.g., Mesarovic, 1964). In the area of applied sciences, General Systems Engineering (GSE) fulfils the role of integrating various "partial" Systems Engineering (SE), which can be divided into three main groups: complexity-related SE, discipline-oriented SE, and

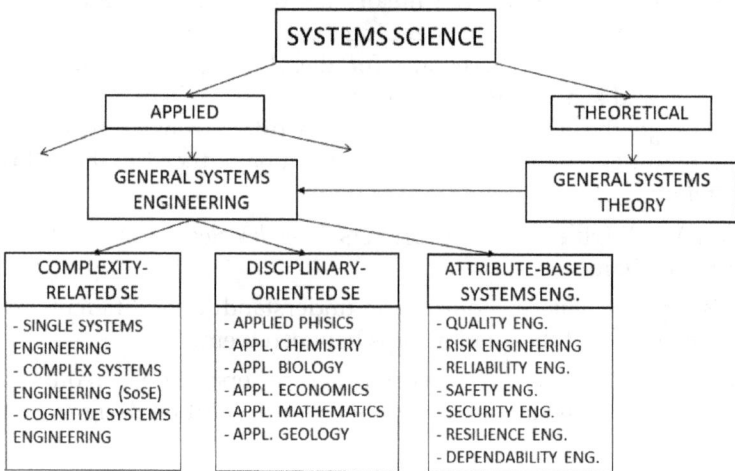

**Figure 3.1:** Hierarchical structure of systems engineering concepts.

attribute-based SE. According to the complexity criterion, SEs can be divided into Single Systems Engineering, Complex Systems Engineering (in practice known as System of Systems Engineering – see Chapter 4), and Cognitive Systems Engineering (see Chapter 5). Selected, disciplinary-oriented systems engineering, and its application examples are shown in Table 3.1.

**Table 3.1:** Selected systems engineering disciplines and its application examples.

| Science Discipline | Engineering Discipline | Application Examples |
|---|---|---|
| Applied physics | Civil Engineering<br>Electrical Engineering<br>Mechanical Engineering | Buildings, bridges, roads<br>generators, electricity grids<br>machinery, motor vehicles |
| Applied chemistry | Chemical engineering | Pharmacy, artificial fertilizers |
| Applied biology | Agricultural engineering | Cultivation of plants and crops |
| Applied mathematics | Software engineering | Computer programs, websites |
| Economics | Financial engineering | Valuation of securities, derivatives |
| Geology | Geological engineering | Mining, drilling platforms |

Attribute-based systems engineering concepts relating to uncertainty will be discussed in detail in Part II (Chapters 6 to 11).

## 3.2 The System Engineering Process – The Lifecycle Concept

In the definition of General Systems Engineering proposed above, a lifecycle orientation is emphasized as a distinctive feature of this approach. The lifecycle concept is the foundation for the analysis of the whole process of system tasks. This process can be divided into 13 stages (Blanchard and Blyler, 2016) and presented as an iterative chain of individual activities related to successive stages of the system's life (see Figure 3.2).

Step 1 requires defining, both qualitatively and quantitatively, the problem, that is, the subject of the project process. All identified relevant needs should be included in the description. In addition to a detailed description of the problem itself, all possible threats and risks that may arise during the implementation of the project should also be considered at this stage. Step 1 is particularly important, because mistakes made at this stage will propagate into the whole chain of activities in the following steps. An additional difficulty is that in the preliminary stages knowledge about the problem to be solved is generally incomplete and subject to the greatest uncertainty. Therefore, it is recommended to consistently apply the iterative procedure in all subsequent steps, including the return to Step 1.

**Figure 3.2:** Life-cycle model of the system engineering process (based on Blanchard and Blyler, 2016).

For the specification of the most important needs, a questionnaire with the following questions can be helpful (Blanchard, 2008):

- What is required of the system, stated in functional terms?
- What specific functions must the system accomplish?
- What are the primary functions to be accomplished?
- What are the secondary functions to be accomplished?
- What must be accomplished to completely alleviate the stated deficiency?
- Why must these functions be accomplished?
- When must these functions be accomplished?
- Where is this to be accomplished, and for how long?
- How many times must these functions be accomplished?

Step 2 is a systematic analysis of the feasibility of the problem to be solved. Once the needs have been established and the problem to be solved has been defined, it should be explicitly checked whether they are feasible. For this purpose, it is necessary to review the various methods, approaches, procedures, and technologies that are available and reasonable to realize the problem defined in the previous step. The following sequence of actions is recommended (Blanchard, 2008):

- identify the various possible design approaches that can be pursued to meet the requirements;
- evaluate the most likely candidates in terms of performance, effectiveness, logistics and maintenance support requirements, and life-cycle economic criteria; and
- recommend a preferred approach.

The aim is to identify and select a general technical and managerial approach to the future system design, without taking the components of the overall system into account at this stage. In a situation where there are many possibilities to realize a project, one should try to limit the number of options to a few options, which are characterized by the highest degree of compatibility with the assumptions and availability of necessary resources.

The next step is to analyse the operational requirements of the system (Step 3). The objective is to reflect the needs of the user in terms of system deployment, utilization, effectiveness, and the accomplishment of its intended functions. The operational concept includes the following general information (Blanchard and Blyler, 2016):

1. Operational distribution or deployment – the number of places where the system will be operated, the geographical distribution and time frame, as well as the type and quantity of major system components in each of these places.

2. Mission profiles or scenarios – specific functions which must the system accomplish to fulfil the required needs.

3. Performance and related operational parameters – clarification of basic operating characteristics of the system expressed in quantitative terms. This applies particularly to characteristics such as speed, rate, capacity, throughput, power output, message clarity, size, weight, operational availability, measure of effectiveness, as well as their range and accuracy.

4. Utilization requirements – expected usage of the system and its elements in accomplishing its functions, e.g., hours of system operation per day, duty cycle, on-off cycles per month, percentage of total capacity utilized, facility loading.

5. Effectiveness requirements – specified quantitatively factors such as system effectiveness and efficiency, operational availability, system reliability and maintainability (e.g., failure rate, mean time between maintenance, maintenance downtime), facility utilization, logistics delay time, lifecycle cost, personnel skill levels, security level.

6. Major system interface or interoperability requirements – the primary interfaces with other systems in a whole hierarchical structure, and the performance requirements that this system must deliver as an input to, or receive as an output from, other systems in a complex structure.

7. Environment definition – identification of the environment in which the system is anticipated to operate in an effective and efficient manner (e.g., temperature range, shock and vibration, noise, humidity).

Step 4 covers the logistics and maintenance support concept and should include the choice of maintenance and renewal policy for the system and its parts. It is proposed to select a *Maintenance Policy* (MP) based on the analysis of seven basic criteria, namely, diagnosing, forecasting, planning, policy, complexity, uncertainty. Table 3.2 summarizes the classification of MP policies according to these criteria.

**Table 3.2:** Classification of maintenance policies according to seven main criteria.

| Criterion | Dichotomous Division | |
|---|---|---|
| Diagnosing | during operation (e.g., Condition-Based M. – CBM) | during machine downtime (e.g., First-Line Maintenance – FLM) |
| Forecasting | probabilistic (e.g., Predictive Maintenance – PM) | deterministic (e.g., Shutdown Maintenance – SM) |
| Planning | Scheduled (e.g., Time-Based Maintenance – TBM) | unscheduled (e.g., Breakdown Maintenance – BM) |
| Policy | model-oriented (e.g., Pro-Active Maintenance – PAM) | potential-oriented (e.g., Emergency Maintenance – EM) |
| Complexity | low or medium (e.g., production systems) | high or very high (e.g., infrastructure networks) |
| Uncertainty | limited (low & medium) | deep (high & very high) |
| Impact | non-critical | critical |

As can be seen from Table 3.2, there are eight main maintenance policies: Condition-Based Maintenance (CBM), First-Line Maintenance (FLM), Predictive Maintenance (PM), Shutdown Maintenance (SM), Time-Based Maintenance (TBM), Breakdown Maintenance (BM), Pro-Active Maintenance (PAM), and Emergency Maintenance (EM). These policies can be classified in practice into three main groups:

(a) 'Fail and fix' procedure (e.g., Breakdown Maintenance),

(b) 'Predict and prevent' strategy (e.g., Predictive Maintenance), and

(c) 'Monitor and self-maintain' policy (e.g., Pro-Active Maintenance – so-called 'Worry-free Performance').

Strategy (a) is only applied when failure of components does not have a significant impact on the continuity and operational safety of the overall system. Strategy (b) is commonly used, e.g., in the Reliability Centreed Maintenance (RCM) system, whereas strategy (c) is used in the Resilience-Based Maintenance (RBM) system and should be implemented for complex systems operating under conditions of deep uncertainty and critical impact of disruptions (see Bukowski and Werbińska-Wojciechowska, 2021). Strategies (b) and (c) can be defined as follows:

- *Reliability Centered Maintenance* – model-oriented policy for systems operating under conditions of limited uncertainty.
- *Resilience-Based Maintenance* – potential-oriented policy for highly complex systems (SoS) operating under conditions of deep uncertainty and critical impact of disruptions.

The correct choice of an appropriate maintenance policy at this early stage of the design process is of particular importance in cases where operational and maintenance costs are significant in relation to the costs of designing and manufacturing the system. Next, organizational responsibility must be clearly defined throughout the operational process of using the designed system. The responsibilities may vary with different components of the system, over time, through operational use of the system and the sustaining support phase. It includes supply support (e.g., spare and repair parts, associated inventories, provisioning data), test equipment, personnel training, transportation, facilities, information flow, and cyber resources.

Subsequently, requirements refer to the logistics and maintenance support infrastructure must be properly integrated with the effectiveness factors associated with system operational requirements. It includes such factors as the availability of the overall support infrastructure, logistics delay time, maintenance downtime, lifecycle cost associated with the logistics and maintenance support capability. The system environment should then be defined by setting parameters such as temperature, shock and vibration, humidity, noise, as a specification necessary for maintenance activities, and related to transportation, handling, and storage functions. To sum up, the logistics and maintenance support concept provides the foundation for establishment of supportability requirements as an input to the system design process and should also provide guidance in the design and procurement of the logistic support elements.

In Step 5, the Technical Performance Measures (TPMs) must be established. There may be some number of goals, initially expressed in general qualitative terms, which the designed system must fulfil in an

**Table 3.3:** An example of quantity requirements for selected Technical Performance Measures.

| Parameters | Numerical Values |
|---|---|
| Operational availability of the system ($A_o$) | > 99% |
| Mean time between maintenance of the system (MTBM) | > 2 000 hours |
| Mean maintenance downtime of the system (MDT) | < 2 hours |
| Mean time between failure for elements (MTBF) | > 5 000 hours |
| Mean corrective maintenance time for software | < 30 minutes |
| Time to identify a fault in a cyber system | < 15 seconds |
| Response time for the logistics and maintenance support | < 24 hours |
| Processing of a purchase order for the acquisition of a component | < 45 minutes |
| Lead time for logistics and maintenance | < 24 hours |
| Critical data access time | < 1 minutes |
| Human operator error rate | < 1% |
| Utilization of the facility | > 95% |
| Unit lifecycle cost for the system | X $ per year |
| Percentage of non-compliant products in total production | < 0,5% |
| Percentage of recycled materials (non-conforming products and used spare parts) | > 98% |

efficient and effective manner. An example of such quantity requirements for selected Technical Performance Measures is shown in Table 3.3. This example relates to a complex cyber-physical system for which operational readiness and reliability requirements are of particular importance. It is above all important to integrate and to view these requirements in the context of the whole. This means that there may be some conflicting issues that will require adjustment when they are considered in the context of the overall system.

Conducting a functional analysis of the system is the task of Step 6. The functional analysis is an iterative process of decomposing system functions into constituent functions performed by the subsystems and their parts, considering all realistic constraints. This analysis can be facilitated through the application of functional-flow block diagrams. Within this procedure the top-level requirements are identified, partitioned to a second level, and on down to the depth. The functional approach helps to ensure the following (Blanchard and Blyler, 2016):

- All aspects of system design and development, production, operation, support, and retirement are covered; that is, all significant activities within the system lifecycle.

- All components of the system are fully recognized and defined, that is, prime equipment, spare and repair parts, test and support equipment, facilities, personnel, data, and software.
- A means is provided for relating system packaging concepts and support requirements to specific system functions, that is, satisfying the requirements of good functional design.
- The proper sequences of activity and design relationships are established, along with critical design interfaces.

The main objective of functional analysis is to ensure *traceability* from the top system-level requirements down to the requirements for detail design.

In practice, a helpful tool for functional analysis is a *functional flow block diagram* (FFBD). The FFBD is a multitier, time-sequenced, step-by-step flow diagram of a system's behaviour. Functional flow relates to the sequencing of operations, with flow arrows expressing dependence on the success of prior operations described by functional blocks. The most important attributes of the method include:

- *Function block* – each function on an FFBD should be separate and be represented by a single box, and needs to stand for a definite, discrete action to be accomplished by system elements.
- *Function numbering* – each level should have a consistent number scheme and provide information concerning function origin. These numbers establish identification and relationships of activities and facilitate traceability from lower to top levels.
- *Functional reference* – each diagram should contain a reference to other functional diagrams by using a functional reference (e.g., box in brackets).
- *Flow connection* – lines connecting functions should only indicate function flow, independent of time.
- *Flow direction* – the flow direction is generally from left to right and arrows are often used to indicate functional flows.
- *Summing gates* – a circle is used to denote a logical gate (AND, OR). AND is used to indicate parallel functions and OR is used to indicate that alternative paths can be satisfied to proceed.

One simple example of a functional flow block diagram is shown in Figure 3.3.

Step 7 concerns the allocation of system requirements. This is a top-down distribution process, which is iterative and often evolving from the results from trade-offs conducted of system components.

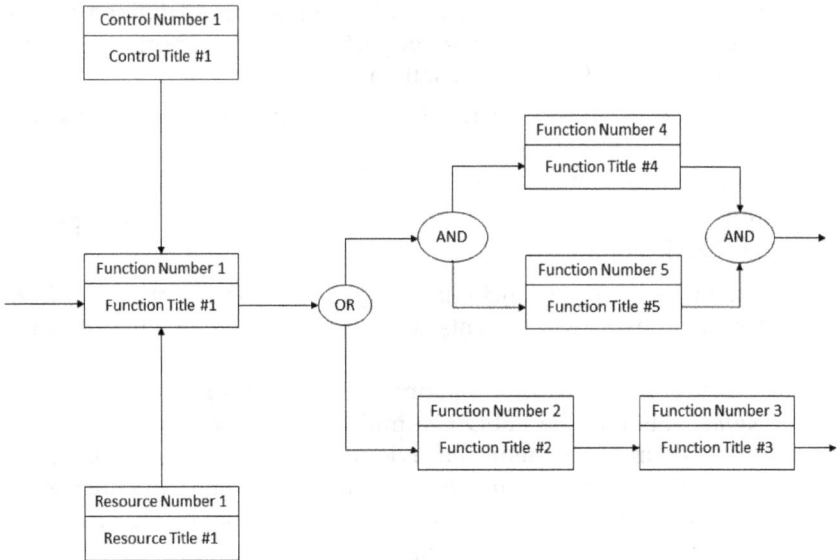

**Figure 3.3:** A simple example of a functional flow block diagram.

The crucial objective is to define specific qualitative and quantitative design requirements for each significant element of the system, and to include such requirements in the appropriate specification for use in the procurement and acquisition process. The basic questions at this stage are as follows (Blanchard, 2008):

- What cyber equipment should be selected that will perform multiple functions?

- How can new functional capabilities be added in the future without adding any new physical elements to the system structure?

- Can supply of any physical resources, e.g., equipment, software, facilities, people, be deleted without losing any of the required functional capabilities previously defined?

A general objective is to break the system down into elements so that only a small number of critical events can influence the functionality of the system. With the identification of system elements, the next step is to allocate the requirements specified for the system down to the level desired to provide a significant input to design. The main challenge is to assign the appropriate factors at the unit level, considering complexity and utilizing historical experience and field data when available. Subsequently, synthesise these factors at the unit level and verify whether they are realistic. There may be both a top-down and a horizontal process in which trade-off studies are accomplished in arriving at a final recommended solution. Sometimes it is necessary to carry out several iterations of this process before the specific requirements for the applicable major system elements are defined.

The analysis and synthesis of the system is addressed in Step 8. The previous steps mainly concerned analysis, while synthesis is a process which refers to the combining and structuring of components in such a way as to represent a feasible system configuration. The synthesis process leads to the definition of several possible alternative design approaches, which will be the subject of further analysis, evaluation, refinement, and optimization. The evaluation procedure progresses through the 10 main steps described as follows (based on Blanchard and Blyler, 2016):

1. Defining the analysis goals. This step requires the clarification of objectives, the identification of possible alternative solutions of the problem, and a description of the analysis approach to be employed. Relative to alternatives, all possible options must be initially considered, but then reduced to a possibly small number of the most promising ones because the bigger number of alternatives considered, the more complex the analysis process becomes.

2. Selecting and weighting evaluation parameters. The basic criteria used in the evaluation process may be differentiated, depending on the stated problem, the specificities of the system under consideration, and the complexity of the analysis. The evaluation parameters of key significance include cost, effectiveness, efficiency, performance, and availability. In any case, parameters are selected, weighted in terms of priority of importance, matched to the system under consideration.

3. Identifying data needs. It is necessary to consider operational requirements, the maintenance concept, major design features, production and/or construction plans, and anticipated system utilization and product support requirements. Fulfilling these needs requires a diversity of data, the scope of which depends on the type of evaluation being performed and the process stage during which the evaluation is accomplished. In the early stages of system development, available data is usually limited. As the system development progresses, more reliable data is available. It is important in the beginning to determine the specific needs for data, and to identify possible data sources. The credibility of the data input can have a significant impact on the risks associated with the decisions made.

4. Identifying evaluation techniques. It is necessary to determine the analytical approach to be used and the techniques that can be applied to facilitate the problem-solving process. Assessing the problem

and identifying the available tools which can be used in solving the problem are necessary prerequisites to the selection of a suitable model.

5. Choosing the right model. A model aids in the development of a simplified representation of the real world as it applies to the problem being solved. The proper model should represent the dynamics of the system configuration being evaluated, highlight the most relevant factors, be comprehensive by including all relevant factors and be reliable in terms of repeatability of results, be simple enough in structure to enable its timely implementation, and be designed to incorporate provisions for easy modification as required.

6. Generating data and model application. The next step is to verify the model to ensure that it is responsive to the analysis requirement (e.g., does the model meet the stated objectives, and is it sensitive to the major parameters of the system configurations being evaluated). Information on evaluating the model and assessing its quality will be presented in Chapter 12, entitled "Methodology of modelling and simulation used for complex systems".

7. Evaluating design alternatives. Several different solutions to a problem can result from modelling studies. Their evaluation and selection of the best option should be based on reliable data. Moreover, the data should be applied in a consistent manner, and the results evaluated in terms of the primarily specified requirements for the system. Only those options that are considered feasible should be selected for further analysis.

8. Accomplishing a sensitivity analysis. The sensitivity analysis of the model should answer the following question: How sensitive are the results of the analysis to possible variations of the input parameters? The main objective of sensitivity analysis is to investigate the relationship between changes resulting from design decisions and their effects.

9. Identifying risks and uncertainty. The identification of all relevant sources of risk and uncertainty in the assessment of model parameters is a very important part of the whole design process. Therefore, these problems are discussed in detail in the following part of the paper, especially in Chapter 16.

10. Recommendation of preferred solution. The final step in the evaluation process is the endorsement of a chosen alternative. All results of the analysis must be fully documented and accessible to all project design participants. A setting of assumptions, a description of the evaluation procedure, an explanation of the various options that were

considered, and an identification of potential risks and uncertainty should be included in this report.

The next step in the lifecycle model (Step 9) is system design integration. This stage requires an integrated team approach, commencing during the early steps of conceptual design and extending through detail system design and development, production, system operation and sustaining support, and ultimate system retirement (recycling or disposal materials). The personnel selected should have the appropriate backgrounds, and can address the 'big picture' approach, as well as can effectively work together and communicate on a regular basis. The main objective is to ensure that the adequate specialists are available at the time required, and that their individual contributions are correctly integrated into the whole.

Step 10 concerns the system evaluation. In practice, a comprehensive evaluation of the system to verify that it meets the initially specified requirements, cannot be carried out until the system is produced and functioning in an operational environment. Nevertheless, if problems occur and system modifications are necessary, this can result in high losses of both time and money. It is therefore advisable to carry out a preliminary evaluation of the system already at this stage, being aware that the outcome of the evaluation may differ from the final assessment. This step can be repeated iteratively as each subsequent step is completed, and the evaluation results should increasingly converge with the final outcome.

In Step 11 the implementation of the system should take place. In most practical applications, these are production or service systems, the specifics of which require the application of methods from a particular area of knowledge. Since there is a very rich literature both in industrial engineering (e.g., Badiru, 2005) and service engineering (e.g., Salvendy and Karwowski, 2010; Bukowski, 2019), this problem will not be discussed further in this paper.

Step 12 concerns the system usage, which means in practice system operational use and sustaining support. A key System Engineering objective during the system operational use phase is to ensure that the system continues to perform as desired by the user. The achievement of this objective is dependent on the sustaining maintenance and support, as well as the availability and implementation of a credible data collection. A properly designed system should also allow for modifications and improvements during its lifetime. However, it should be borne in mind that any proposed system modification must be evaluated from a total life-cycle perspective, and a detailed plan for incorporation must be prepared and accurately implemented.

The last step (Step 13) refers to system retirement and material recycling or disposal. To limit the negative impact of system decommissioning on the environment, the requirements for disposal conditions of end-of-life systems and their components should also be considered. The residual material must be processed through some form of disposal. So, the systems engineering process should include perspectives such as design for supportability, design for disposability, and design for the environment.

In summary, it should be stated that only a holistic and process-oriented approach to Systems Engineering can ensure that the designed system meets all requirements, namely, functional as well as cost and environmental.

## 3.3 References

Badiru, A. (ed.). (2005). *Handbook of Industrial and Systems Engineering.* CRC Press.

Blanchard, B.S. (2008). *System Engineering Management.* NY: Wiley.

Blanchard, B.S. and Blyler, J.E. (2016). *System Engineering Management.* NY: Wiley.

Bukowski, L. (2019). *Reliable, Secure and Resilient Logistics Networks. Delivering Products in a Risky Environment,* Springer Nature Switzerland AG 2019. ISBN 978-3-030-00849-9 (Hardcover), ISBN 978-3-030-00850-5 (eBook).

Bukowski, L. and Werbińska-Wojciechowska, S. (2021). *Using Fuzzy Logic to Support Maintenance Decisions according to Resilience-Based Maintenance Concept, Eksploatacja I Niezawodność – Maintenance and Reliability.* ISSN 1507-2711; 2021; 23 (2): 294–307. http://doi.org/10.17531/ein.2021.2.9.

DAG. (2013). *The Defense Acquisition Guidebook.* https://at.dod.mil/sites/default/files/documents/DefenseAcquisitionGuidebook.pdf .

DOD. (2002). Department of Defense Regulation 5000.2R. *Mandatory Procedures for Major Defense Acquisition Programs (MDAPS) and Major Automated Information Systems (MAIS) Acquisition Programs.*

INCOSE. (2011). *Systems Engineering Handbook: A Guide for System Life Cycle Processes and Activities,* Version 3.2.2. San Diego, CA: INCOSE.

Mesarovic, M.D. (1964). *Foundation for Mathematical Theory of General Systems.* NY: Wiley.

Salvendy, G. and Karwowski, W. (eds.). (2010). *Introduction to Service Engineering.* NY: Wiley.

# 4

# System of Systems Engineering[*]

The chapter deals with the further development of Systems Engineering into more complex systems, namely System of Systems Engineering. The basic properties distinguishing complex systems are first discussed, especially the concepts of synergy and multidimensional complexity. Different aspects of complexity are analysed, such as structural, spatial, temporal, and disciplinary. On this basis, the concept of System of Systems (SoS) is introduced as a mix of independently operating and actively interacting large systems, integrated with complex goals. A general mathematical model for SoS-type systems and detailed models for network structures, typical of complex network organizations, have been proposed. The following part of the chapter is devoted to Engineered System of Systems and particularly to the process of creating such systems, i.e., architecting. The purpose of the architecting process is to provide the required properties to the created systems by incorporating into it the so-called *metasystem* that fulfils a governance role. The functional structure of the metasystem is proposed and the basic functions and sub-functions that a properly designed metasystem should fulfil are characterized.

## 4.1 Main Principles of Complex Systems Engineering

The system approach gained significance in recent decades, due to the growing number of the so-called *large-scale problems*. These include important issues in both defense and civilian systems, in terms of production and services, and examined from different perspectives – technical, economic, social, environmental, and political. The root of these problems lies in complex interactions between the elements and

---

* https://orcid.org/0000-0002-2630-3507.

states of the system, as well as unpredictable changes in the environment caused by forces of nature (e.g., natural disasters) or intended actions (e.g., criminal and terrorist activities). Large-scale problems lead to the phenomenon of so-called *mess*, defined by Russell L. Ackoff as "a state of an organization (or any other system) caused by a combination of unexpected and interrelated situations that pose a threat or an opportunity to the organization or system" (Ackoff et al., 2006). A lack of appropriate systemic response to the mess can lead to a deep crisis or even destruction and collapse of the entire organization (system). Traditional problem-solving methods are generally ineffective, which may be due to the so-called circular logic effect. This effect lies in the fact that in complex systems with many interactions between elements and state variables, certain system states can be both the causes and effects of changes in the system. This is one of the main reasons causing difficulties in identifying both the problem itself and the causes that could lead to it.

The practical implementation of the idea of a systems approach for complex systems has become possible thanks to the dynamic development of information technology. It requires a synergetic cooperation in numerous areas using a variety of methods and tools, e.g., in accordance with the so-called $C^4I^2$ principle, whose name is an acronym for the following terms: Command, Control, Communication, Computer, Integration, Intelligence. The main objective of this rule is to support decision-makers in handling unexpected situations, which they have never had to deal with (Skyttner, 2008).

One of the most important criteria of real systems division is the degree of their complexity (see Table 2.2). In general, it can be assumed that there are three levels of complexity in systems: *single systems* (simple or complicated), *multiple, complex systems*, often referred to as *systems of systems* (Gideon et al., 2005), and the most complicated *cognitive systems*. The main difference between these classes of systems is that in the case of single systems, their properties are determined by the properties of their components and the nature of internal relations between these elements, whereas the properties of complex systems as well as cognitive systems are *emergent* depending on internal and external conditions (Eusgeld et al., 2011). Cognitive systems will be discussed in Chapter 5, while the subject of this section is System of Systems (SoS).

The concept of *emergence* comes from the Latin word 'emergo', which means 'to arise'. It denotes the phenomenon of the generation of qualitatively new structures and behaviours because of complex interactions between many elements in the system and complex forces coming from the system's environment. This concept is crucial in the description of relevant properties of all complex systems and is based on the so-called '6 N' rule which refers to features such as novelty, nonreducibility, non-deducibility, non-predictability, non-computability, non-explainability. In line with J.S. Mill (2002), "emergence is the inability to designate a certain value related to the total result of causes operating together, as a resultant of relevant values related to individual causes, based on a specific rule for composing given values". The conditions that a system must fulfil to be considered emergent are:

- the impossibility of explaining its properties and behaviours based on the knowledge about the components of a complex structure, which they form (so-called *epistemic emergence*),
- lack of possibility of clarifying its properties and behaviours based on the knowledge about the relationships between the components (so-called *interactive emergence*),
- inability to predict the existence of the emergent property when having knowledge of the characteristics of these components (so-called *actualizing emergence*).

In practice, emergence refers to phenomena observed in *big-picture scale*, rather than occurring in small-picture scale. The typical examples of emergent systems include living organisms such as ant colonies, termite mound, swarms of bees, bird V formation, schools of fish, flocks of wolves, and even crowds of people.

The *complexity* of systems is multidimensional in its nature, namely, structural, spatial, temporal, and disciplinary. *Structural complexity* is caused by the big number of elements that make up the system, the quantity of different types of elements that form the structure of the system, and the number of active connections between the elements of the system. However, there are no clear quantitative models to unambiguously distinguish between simple and complex systems and determine the boundaries between these systems. In practice, it is assumed that at least one of these three conditions must be met for a system to be classified as a complex system.

*Spatial complexity* is related with the scope the system covers. For organization type systems, a division can be made into small-scale systems (e.g., covering the scope of a single company), large-scale systems (e.g., regional supply networks), very large-scale systems (e.g., national critical infrastructure), and ultra-large-scale systems (e.g., global supply networks). *Time related complexity* results from the variability in the system's behaviour over time. Static and quasi-static systems represent a low level of complexity. A higher level of time complexity is typical for dynamic systems of regular variation (e.g., periodic), and the highest level of temporal complexity is characteristic of irregularly variable dynamic systems (e.g., random, or chaotic).

The disciplinary division of systems is hierarchical and the following four levels are distinguished:

- systems within a single discipline – monodisciplinary systems,
- systems within many scientific disciplines – multidisciplinary systems,
- systems on the border of defined scientific disciplines – interdisciplinary systems, and
- systems that exceed all boundaries between disciplines and combine all the above features in an inclusive way – transdisciplinary (or over-disciplinary) systems.

## 4.2  System of Systems and Network Organization

Real complex systems are usually characterized by nonlinear structures (e.g., networks), spatial scope (e.g., at least a large scale), dynamic behaviour (e.g., responsive, or active), and exceeding a single scientific discipline (interdisciplinary or transdisciplinary approach). Because research methods used in classical systems engineering are insufficient for the analysis, synthesis, and management of these types of systems, the so-called System of Systems Engineering has been developing for several years in parallel to Single Systems Engineering. Table 4.1 shows the main differences between these two areas of knowledge. From a practical point of view, it is very important that one should pay attention to the multiplicity of goals that occur in complex systems, limited capabilities of

**Table 4.1:** Comparing the main attributes of System Engineering and System of Systems Engineering.

| Attribute | System Engineering | System of Systems Engineering |
|---|---|---|
| Focus | single complex system | multiple integrated complex systems |
| Problem | defined | emergent |
| Objective | optimization (best) | satisfying (good enough) |
| Boundaries | static (fixed) | dynamic (variable) |
| Structure | hierarchy or chain | network |
| Goal | unitary | pluralistic |
| Approach | system related | methodology related |
| Timeframe | life cycle | continuous |
| Design tools | many | few |
| Standards | few | non |

*Source*: Based on Jasmhidi, 2011; Souza-Poza et al., 2008; Valerdi et al., 2008; Zio, 2007.

finding solutions within these systems (searching for satisfactory solutions instead of optimization) and extending the time horizon in the research of complex systems from the scope of a life-cycle to a continuous, limitless one (as a result of the pursuit of sustainability).

SoS, as a natural extension of large-scale systems (LSS), represents a mix of independently operating and actively interacting large systems, integrated with complex goals. So far it has not been possible to unambiguously define the term SoS, so it is generally described by the features that distinguish this class of systems from others. These features include (Bukowski 2019):

- multidimensional complexity – nonlinear and heterogeneous structures (e.g., networks), spatial scope of at least a large scale, dynamic behaviour, and going beyond a single scientific discipline (interdisciplinary or transdisciplinary approach),
- operational and managerial independence of its elements (subsystems) – the subsystems must be able to usefully operate independently and maintain a continuing operational existence independent of the SoS,
- emergent behaviour – the SoS performs functions and carries out purposes that do not reside in any subsystem, and
- evolutionary development – development and existence of SoS is evolutionary with functions and purposes added, removed, and modified with experience.

In industrial and service practice, network organizations are a typical example of application of SoS-type structures. *Network organization* is a structurally, spatially, and temporally complex process organization, whose elements comprise of independent enterprises, and its properties are emergent in their nature. The main features of network organizations include (based on Bukowski, 2019):

- Flat structure with the topology of a weighted digraph, with fuzzy borders between the organization and its environment.
- Spatial scope reflecting the geographical scale of the organization extending beyond one company.
- Temporal variability resulting from environmental dynamics and the shortening of the life cycle of products and services.
- The capacity to change expressed by flexibility and agility, resulting from the functional redundancy (simultaneous execution of processes).

- Synergy resulting from the full compatibility of competences of units forming the network organization, the adoption of common goals, and the organizational capacity to learn.
- Organizational intellectual capacity through knowledge sharing between the units of the network and the accumulation of emerging knowledge.

General mathematical model of such complex systems can be described as following (Bukowski 2019):

$$S = \{S_1, \ldots, S_i, \ldots, S_k\} \tag{4.1}$$

$$S = (S \times \mathcal{R}) \tag{4.2}$$

where:  $S$ – modelled complex system (e.g., System of Systems),
$S$ – set of subsystems,
$S_i$ – subsystem $i$ (element of the complex system),
$k$ – number of subsystems,
$\times$ – Cartesian product,
$\mathcal{R}$ – relations between subsystems $S$.

The structure of complex networks can be described using graphs in the following form (Newman, 2010):

$$G = (V, E) \tag{4.3}$$

where:  $G$ – directed or undirected graph,
$V$ – vertex (set of nodes in network), and
$E$ – edge (set of lines connecting two nodes).

Figure 4.1 shows examples of the most common network topologies: (a) Bus; (b) Hierarchical; (c) Ring; (d) Star; (e) Mesh; (f) Linear; (g) Hybrid. In this figure, the black points represent the *nodes* of the network, while the lines connecting these points are *edges*. In accordance with the mathematical theory of graphs, the basic size describing the network node is the so-called *node degree d*. It expresses how many ends of lines belong to a given node, i.e., how many of the given nodes have partners or neighbours. The frequency of nodes occurrence with various node degrees can be described by the binomial model (also called the Bernoulli distribution), which characterizes a network with $N$ random connected nodes. These networks are called *random networks* of the ER-type (Erdos-Renyi model). Since this type of network rarely occurs in the real world, random networks described by a power-law

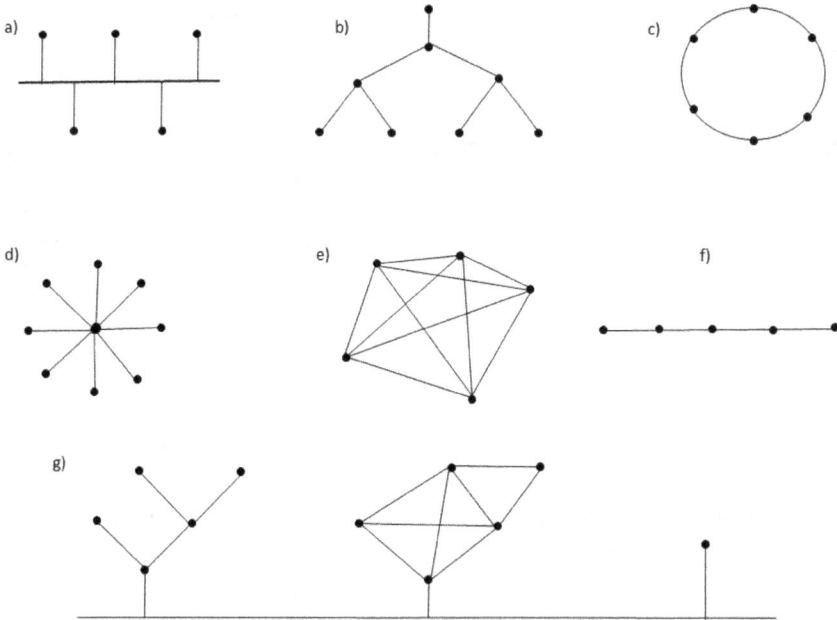

**Figure 4.1:** The most common network topologies: (a) Bus; (b) Hierarchical; (c) Ring; (d) Star; (e) Mesh; (f) Linear; (g) Hybrid (combination of Bus, Hierarchical, and Mesh topologies).

distribution are generally used (also called Pareto-law) with the following formula:

$$P(d) = a \cdot d^{-\alpha} \tag{4.4}$$

where:  $P$ – probability of $d$,

$d$ – node degree,

$a$ – constant value,

$\alpha$ – exponent's value (a number within the range 2 to 4).

Networks whose structure is described by the above model are called *scale-free networks* (Barabasi et al., 1999), and their main feature is that they are invariant to the operation of scaling the degree of nodes. This means that changing the degree of the nodes does not affect the probability distribution, and it continues to follow a power-law distribution. The topology of the scale-free network is characterized by a small number of nodes with high degrees $d$, so-called hubs, and a relatively large number of nodes with low degrees $d$. It results in such a network that is resistant to accidental random failures but susceptible to deterministic faults (e.g., precisely planned attacks). In practice, there are many modifications of these types of network structures, and one of the most popular is the so-called *small-world network* (Watts and Strogatz, 1998). A small-world network is a system in which the neighbours of any given node are likely to be neighbours of each other and most nodes can be reached from every other node by a small number of steps. The effect of this property is that the average distance $L$ between two randomly chosen nodes (the number of steps required) grows proportionally to the logarithm of the number of nodes $N$ in the network.

All types of networks can be characterized by defining basic measures for their properties. One of the most important is identification of central units in a network – centrality characteristics. *Measures of centrality* can be defined in two different ways (Newman, 2010):

- for each unit separately, as a unit centrality (one number for each unit), or
- for the whole network, as a network centralization (only one number for the whole network).

Describing the *network centrality* should be distinguished between undirected and directed networks:

- the term, *centrality measures*, is used for undirected networks, which means that a communication node is central, if a lot of roads are passing through it,
- the term, *prestige*, is used for directed networks. In this case, two different types can be defined: one for outgoing edges (so-called measures of influence), and one for incoming edges (so-called measures of support).

In addition, a distinction is made between:

*Unit Centrality* – selected unit is central, if it has a high degree, is close to all other units, and lies on the shortest paths between other units.

*Degree Centrality* – a unit is central in a network, if it is active enough in the sense that it has a lot of links to other units. In the case of cycle, all units are equally central. Degree centrality is defined as follows:

$$C_D(x) = degree\ of\ unit\ x \qquad (4.5)$$

Such measures are called *absolute measures of centrality*. Absolute measures cannot be used to compare centralities of networks with different number of units. Therefore, they are normalized to get a measure in intervals from 0 to 1, where 0 means the smallest possible and value 1 the highest possible centrality. Measures normalized in this way are called *relative measures of centrality*.

*Closeness Centrality* – the measure of centrality according to closeness of unit $x$:

$$C_C(x) = 1/\Sigma_{y \in U} d(x, y) \qquad (4.6)$$

where $d(x, y)$ is the graph-theoretic distance (length of the shortest path) between units $x$ and $y$, and $U$ is set of all units. If a network is not strongly connected, we take only reachable nodes into account, but we weigh the result with several reachable nodes. The most central units according to closeness centrality can rapidly interact with all others because they are close to all others. This measure considers both direct and indirect connections among units.

*Relative Closeness Centrality* is defined by the formula:

$$C_C(x) = (n - 1) \cdot C_C(x) \qquad (4.7)$$

The smallest possible distance of selected unit from all other units is obtained, if the unit has all other units for neighbors. In this case the absolute closeness centrality is $1/(n - 1)$. Closeness centrality can be calculated for undirected and directed networks. There are two possibilities for directed networks: prestige can be calculated according to outgoing edges (how close are all other units from the selected one; which means in how many steps is needed to reach all other units from the selected one), or according to incoming edges (how close is the selected unit to all others; it corresponds to how many steps is needed to reach the selected one from all other units).

*Betweenness Centrality* – in the case of communication it is very important to identify the nodes which lie on the shortest paths among pairs of other units. Such units have control over the flow of information in the network. The concept of betweenness centrality measures assumes that a unit is central if it lies on several of shortest paths among other pairs of units. We can define the centrality measure of unit $x$ according to betweenness in the following way:

$$C_B(x) = \sum_{y < z} \frac{shortest\ path\ from\ y\ to\ z\ through\ x}{shortest\ path\ from\ y\ to\ z} \qquad (4.8)$$

If communication in a network always passes through the shortest of available paths, then Betweenness Centrality of a unit $x$ is the sum of probabilities across all possible pairs of units, that the shortest path between $y$ and $z$ will pass through the unit $x$.

## 4.3　Architecting Engineered System of Systems

Complex engineering systems, such as a SoS or a network organization, consist of three basic systems:

- infrastructure implementing the required functions (e.g., the network organization),
- protection subsystem (e.g., safety and security barriers), and
- meta-system of management (e.g., the flow of information and communication within the SoS).

Examples of such man-made SoS include power supply networks (so-called smart grids), the Internet of things, telecommunication networks, and global supply chains. They are characterized by the following features: a complex distributed structure, a high degree of autonomy of the individual components, a multi-state dynamics, emergent properties, as well as adaptability to rapidly changing target (e.g., demand-driven supply chain).

*Engineered System of Systems (ESoS)* is a set of heterogeneous man-made subsystems assembled purposefully together to achieve a common goal that any system alone cannot fulfil, while maintaining the operational and managerial autonomy of each of the subsystems (Jamshidi, 2005; Bukowski, 2019). These subsystems must be able to communicate and to work harmoniously together as well as to adapt their behaviour and function locally when facing any change of their environment (Jamshidi, 2011), which means concentrating activities on choosing and assembling the subsystems and designing appropriate interfaces to facilitate the reliable communication between individual parts of the whole system. Subsystems are selected and involved according to their potential roles, available resources, competencies, and know-how that can be shared to fulfil the SoS objectives (Bilal et al., 2014).

The process of building ESoS from subsystems is called *architecting*. The purpose of the architecting process is to provide the required properties to the created systems. The most important required properties are (Billaud et al., 2015):

- *Extensibility* of an open system is its ability to add new components, subsystems, or systems, as well as new capabilities to a system.
- *Flexibility* implies that a given system, depending on the current requirements, can be reconfigured, and modified to varying situations.
- *Integratability* provides the system with the ability to be able to form, coordinate, or incorporate into a larger, functioning whole.
- *Interchangeability* ensures that a given system or a part of it can be replaced with another one without losing the basic system properties and features.

- *Interoperability* means the ability of connected, autonomous, flexible coupled and usually heterogeneous systems to cooperate and to exchange streams of data, services, material, and energy to and from other systems, while continuing their own way of operation.
- *Modularity* of a given system implies that it is built of functional blocks, separating the system's capacities into modules.
- *Portability* is the capacity to be moved from one environment to another.
- *Replaceability* is the ability of a system, component, or person to take the place of another, especially as a substitute or successor.

The subsystems included in the ESoS are heterogeneous and autonomous, however the entire system must achieve objectives which may be significantly different from any of the subsystems. Therefore, each Engineered System of Systems must contain an additional *metasystem* that fulfils a management or governance role in relation to all subsystems of ESoS. Thus, the design, instalment, operation, and transformation of a metasystem play a key role in architecting reliable ESoS. Because every metasystem is comprised of autonomous embedded complex systems, it can diversify in technology, context, operation, geography, and conceptual frame.

The first step in the metasystem architecture process, the creation of the *metasystem construct* (MSC), brings several important considerations, which can be summarized as follows (Keating and Katina, 2016):

- MSC operates at a logical level beyond the system, subsystems, and entities that it must integrate.
- The concept of MSC has been conceptually grounded in the foundations of Systems Theory (axioms and propositions governing system integration and coordination) and Management Cybernetics (design of the communication and control for effective system organization).
- The MSC is equipped with several interrelated functions, which only specify 'what' must be achieved for continuing system existence, not specifying 'how' those functions are to be achieved.
- The most important role of these functions is to ensure the survival of the system, even at the cost of reducing its performance and quality of operation.

The metasystem construct is the basic element of any ESoS and determines its operational capacity. The metasystem is the 'governor' in a cybernetic sense of providing control for an entire system, and this type of control is essential to ensure a system the stability of performance in situations external environmental changes or turbulences. Control generated by the metasystem is achieved in conjunction with three key roles (Keating et al., 2014), including:

- *Communication*. The main tasks of communication are organization of the flow, transduction, and processing of information internal and external to the system.
- *Coordination*. It provides interactions between essential entities within the system, and between the system and external elements to avoid undesirable instabilities and disturbances.
- *Integration*. It ensures continuous maintenance of system integrity, a dynamic balance between autonomy of constituent entities and the integration of those entities to form a coherent whole.

The second step in the architecture process involves the *governance functions*, including four primary functions and five associated sub-functions. These primary functions can be described as follows (Keating et al., 2014):

- *Policy and identity* maintain and define the balance between current and future state of an organization from two main perspectives:
  - system perspective – focused on the specific system context within the metasystem is implanted, and
  - strategic perspective – focused on monitoring of the system performance indicators at a strategic level.
- *System development* concentrates on the long-range development of the system to ensure future viability by:
  - environmental monitoring – supervision of the environment in search of trends, patterns, or untypical events, and
  - learning and transformation – correction of eventual design imperfections in the metasystem functions and communication channels as well as preparation of changes or transformations in the metasystem.
- *System operations* support the current execution of the metasystem to ensure that the overall system maintains required performance levels, and in particular:
  - operational performance – monitoring system performance to identify and assess abnormal conditions, exceeded thresholds, or anomalies.
- *Information and communications* responsible for design, establishing, and maintaining the flow of information through communication channels, and consistent interpretation of exchanges necessary to fulfil metasystem functions.

Table 4.2 shows summary of main characteristics of the communication channels, their primary metasystem function responsibility, and the description of the role each of them play in metasystem execution.

An example of the functional structure of an Engineered System of Systems is shown in Figure 4.2.

**Table 4.2:** The governance functions fulfilled by a metasystem.

| Function | Sub-functions | Description of the Function's Role |
|---|---|---|
| Policy and identity | Command | Provides non-negotiable direction to the metasystem and governed systems |
| | Control | Provides for examination of system decisions, actions, and interpretations for consistency with system purpose and identity |
| | Emergency | Provides redundancies of all channels when the integrity of the system is threatened and compels instant alert to crisis or potentially catastrophic situations for the system |
| System development | Environmental monitoring | Provides design for sensing to monitor critical aspects of the external environment and identifies environmental patterns, activities, or events with system implications |
| | Learning and transformation | Provides detection and correction of error within the metasystem as well as governed systems, focused on system design issues as opposed to execution issues |
| System operations | Resource management | Determines and allocates the resources (manpower, material, money, methods, time, information, support) to governed systems and defines performance levels, e.g., productivity, responsibilities, and accountability for governed systems |
| | Operations management | Provides for the routine interface concerned with near term operational focus; concentrated on providing direction for system production of value (products, services, processes, information) consumed external to the system |
| | Audit | Provides routine and sporadic feedback concerning operational performance as well as investigation and reporting on problematic performance issues within the system |
| Information and communication | Coordination | Provides for metasystem and governed systems balance and stability as well as ensures design and achievement (through execution) of design:<br>• ensuring that decisions and actions necessary to prevent disturbances are shared within the metasystem and governed systems, and<br>• sharing of information within the system necessary to coordinate activities |
| | Informing | Provides for flow and access to routine information within the metasystem or between the metasystem and governed systems |

*Source*: Based on Keating and Katina, 2016.

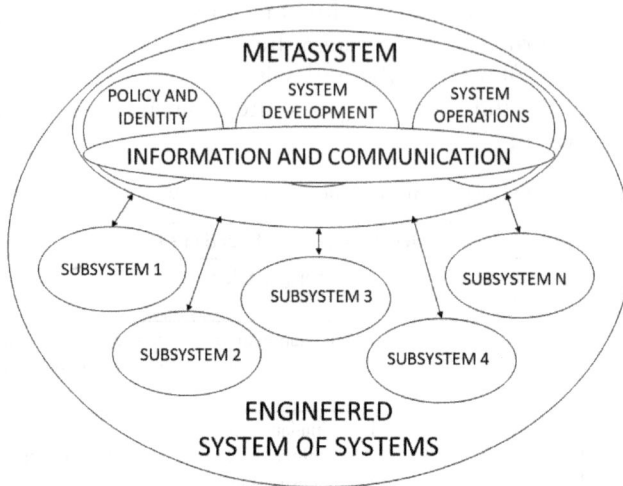

**Figure 4.2:** An example of the functional structure of an Engineered System of Systems.

A metasystem consisting of three separate parts, namely, policy and identity, system development and system operations, is integrated by information and communication functions. Such a functional structure of the ESoS ensures that even very complex systems (e.g., global supply networks) can be fully controlled and properly governed.

## 4.4 References

Ackoff, R.L., Magidson, J. and Addison, H.J. (2006). *Idealized Design: Creating an Organization's Future.* Pearson Education Ltd.

Barabasi, A.-L. et al. (1999). Emergence of scaling in random network. *Science,* 286(5439): 509–512.

Bilal, M., Daclin, N. and Chapurlat, V. (2014). *Collaborative Networked Organizations as System of Systems: A Model-based Engineering Approach.* IFIP AICT.

Billaud, S., Daclin, N. and Chapurlat, V. (2015). Interoperability as a key concept for the control and evolution of the System of Systems (SoS), Conference: *International Workshop on Enterprise Interoperability* (IWEI, 2015). Doi: 10.1007/978-3-662-47157-9-5. https://www.researchgate.net/publication/279928616_Interoperability_as_a_Key_Concept_for_the_Control_and_Evolution_of_the_System_of_Systems_SoS.

Bukowski, L. (2019). *Reliable, Secure and Resilient Logistics Networks. Delivering Products in a Risky Environment.* Springer Nature Switzerland AG 2019. ISBN 978-3-030-00849-9 (Hardcover), ISBN 978-3-030-00850-5 (eBook).

Eusgeld, I., Nan, C. and Dietz, S. (2011). System-of-systems approach for interdependent critical infrastructures. *Reliability Engineering and System Safety,* 96(6): 679–686.

Gideon, J.M., Dagli, C.H. and Miller, A. (2005). Taxonomy of Systems-of-Systems. *Proceedings CSER 2005,* 23–25 March, Hoboken, NJ, USA 363.

Jamshidi, M. (2005). System-of-Systems engineering – A definition. *In: IEEE International Conf. on System, Man and Cybernetics (SMC).* Waikoloa, Hawaii, Vol. 4, 10–12 October 2005. http://ieeesmc2005.unm.edu/SoSE_Defn.htm.

Jamshidi, M. (2011). *System of Systems Engineering: Innovations for the Twenty-first Century.* NY: John Wiley & Sons.

Keating, C. and Katina, P. (2016). Complex system governance development: A first generation methodology. *International Journal of System of Systems Engineering,* 7(1/2/3): 43–74.

Keating, C.B., Katina, P.F. and Bradley, J.M. (2014). Complex system governance: Concept, challenges, and emerging research. *International Journal of System of Systems Engineering,* 5(3): 263–288.

Mill, J.S. (2002). *A System of Logic.* Honolulu: University Press of the Pacific. ISBN 1-4102-0252-6.

Newman, M.E.J. (2010). *Networks. An Introduction.* Oxford University Press.

Skyttner, L. (2008). *General Systems Theory. Problems, Perspectives, Practice.* Singapore: Word Scientific.

Sousa-Poza, A., Kovacic, S. and Keating, C. (2008). System of systems engineering: An emerging multidiscipline. *International Journal of System of Systems Engineering,* 1: 1–17.

Valerdi, R. et al. (2008). A research agenda for system of systems engineering. *International Journal of System of Systems Engineering,* 1: 171–188, Geneva: Interscience Publisher.

Watts, D.J. and Strogatz, S.H. (1998) Collective dynamics of 'small-world' networks. *Nature,* 393: 440–42.

Zio, E. (2007). From complexity science to reliability efficiency: A new way of looking at complex network systems and critical infrastructure. *International Journal of Critical Infrastructure,* 3(3/4): 488–508.

# 5

# Cognitive Systems Engineering[*]

The purpose of Chapter 5 is to define the main features that distinguish Cognitive Systems Engineering from other systems engineering. The first section is devoted to the discussion of basics of Cognitive Science and Cognitive Systems, and in particular its interdisciplinary nature and relationships between cognitive systems and major classes of systems. This is followed by a brief analysis of cognitive processes of knowledge acquiring in living systems as a chain consisting of basic, meta, and higher cognitive functions. Section 3 provides processes of knowledge acquiring in engineered systems which is described and analysed based on a modified knowledge pyramid. In addition, a simplified model of knowledge acquisition process in engineering systems is proposed. The main part of the chapter is Section 4, which defines Cognitive Systems Engineering based on the imperfect knowledge acquisition and management concept. The model of the knowledge extraction process as a chain of actions that must be performed to obtain knowledge from raw data is proposed. The chapter concludes by showing the relationship of Cognitive Systems Engineering to other areas of knowledge and proposing a structural model for describing Cognitive Engineered Systems.

## 5.1 Basics of Cognitive Science and Cognitive Systems

*Cognitive Science* is the interdisciplinary field of knowledge based on studies of mind and intelligence, embracing philosophy, psychology,

---

[*] https://orcid.org/0000-0002-2630-3507.

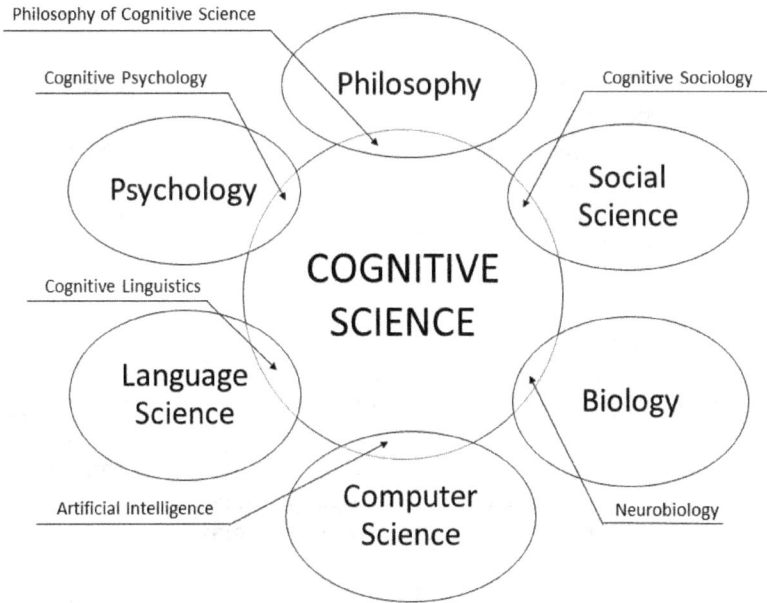

**Figure 5.1:** Cognitive Science as an interdisciplinary field of knowledge (based on https:// plato.stanford.edu/entries/cognitive-science/).

social science, biology, language science, and computer science (see Figure 5.1). The term 'cognitive' in 'cognitive science' is used for "any kind of mental operation or structure that can be studied in precise terms" (Lakoff and Johnson, 1999). This conceptualization is very broad and means roughly pertaining to the action or process of knowing. The result of this interdisciplinarity of research in the field of cognitive science is a great diversity of views and concepts resulting from the variety of methods and research objectives that characterize the different areas of knowledge.

Philosophy of cognitive science deals with general questions such as the relation of mind and body and with methodological questions such as the nature of explanations found in knowledge. This applies particularly to normative issues, e.g., how people should think and with descriptive ones about how they do it. The main theoretical goal of cognitive science is to find a model of human thought processes, while the practical goals come down to finding ways to improve the efficiency of these processes.

It is commonly believed that the other components of the term 'cognitive science' (i.e., psychology, social science, biology, language science, and computer science) should be considered as equally important parts of the whole system, that is, Cognitive Science. This is due to a paradigm called *cognitive commitment*, which assumes that all components of cognitive processes form a coherent whole and are interconnected by strong interactional relationships (Talmy, 2000; Wolfson, 2001).

Starting from this paradigm, it is assumed that the best way to understand the complexity of human thinking is using multiple methods, especially psychological experiments supported by computational models and simulations. Psychologists have experimentally examined the kinds of mistakes people make in deductive reasoning, the ways that people form and apply concepts, the speed of people thinking with mental images, and the performance of people solving problems using analogies. Cognitive theorists representing artificial intelligence have proposed that the mind contains mental representations of logical propositions, rules, concepts, images, and analogies, and that it uses mental procedures such as deduction, search, matching, rotating, and retrieval. Cognitive science then works with an analogy among the mind, the brain, and computers, whereby each of them may be used to propose new ideas about the others.

A major trend in current cognitive science is the increasing integration of neuroscience with many areas of psychology, including cognitive, social, developmental, and clinical, as well as artificial intelligence. This integration is mainly experimental, resulting from the dynamic maturation of new research methods and tools for studying the brain, such as functional magnetic resonance imaging, transcranial magnetic stimulation, and optogenetics. Thanks to this research, there is also significant progress in theoretical foundations of cognitive science, because of advances in understanding how large populations of neurons can perform tasks explained with cognitive theories of rules and concepts. There is also a growing awareness of the importance of the influence of the external environment, both physical and social, on cognitive processes.

In line with the proposed division of systems in Section 2.2, Figure 5.2 illustrates the relationship between these systems and the notion of a cognitive system. This concept implies that cognitive systems are fragments of collections of those systems that satisfy the conditions for cognitive systems. They are, therefore, cognitive living systems, cognitive sociological systems, cognitive natural systems, cognitive artificial systems, cognitive abstract systems, and cognitive ICT systems. In accordance with a simplified division into physical systems (natural and artificial), cyber systems (ICT and abstract) and social systems (living and sociological) can be distinguished: Cognitive Physical Systems, Cognitive Cyber Systems, and Cognitive Social Systems.

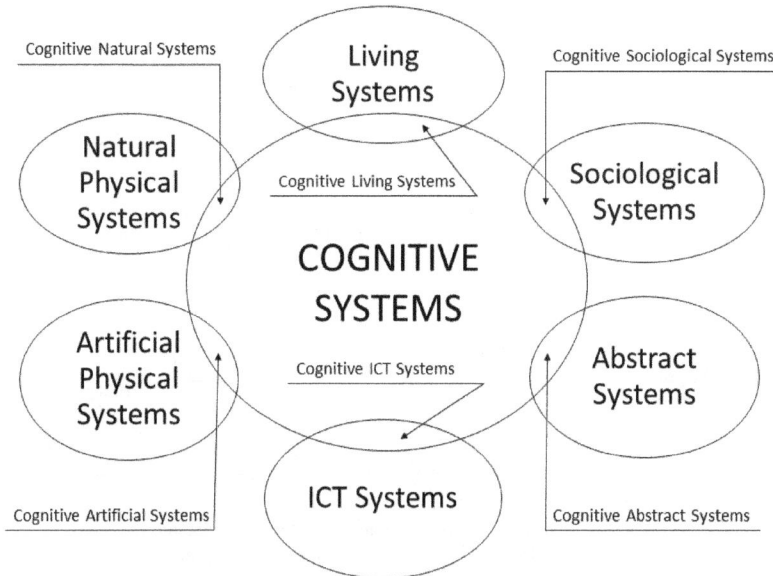

**Figure 5.2:** Relationships between cognitive systems and major classes of systems.

## 5.2 Cognitive Processes of Knowledge Acquiring in Living Systems

*Cognitive process* is defined as any of the mental functions involved in the acquisition, storage, interpretation, manipulation, transformation, and use of knowledge. These processes encompass such activities as attention, perception, learning, and problem solving and are commonly understood through several different theories, as well as the serial processing approach and parallel processing approach. The term 'cognitive processes' is often synonymous with the concept of mental processes (https://dictionary. apa.org/cognitive-process).

Cognitive processes can also be understood as chains consisting of basic, meta, and higher cognitive functions. Figure 5.3 shows a concept of such a chain, consisting of seven cognitive functions connected by serial relationships and feedback. The first function in the chain is *sensation* which occurs when special receptors in the sense organs (e.g., the eyes, ears, nose, skin, or taste buds) are activated, allowing various forms of outside stimuli to become neural signals in the brain. The process of converting outside stimulus into neural activity is called transduction. *Attention* is the behavioural and cognitive function of selectively concentrating on a discrete aspect of stimuli (signals) while ignoring other perceivable inputs. It can also be described as the allocation of limited cognitive processing

Basic cognitive functions                    Meta cognitive functions

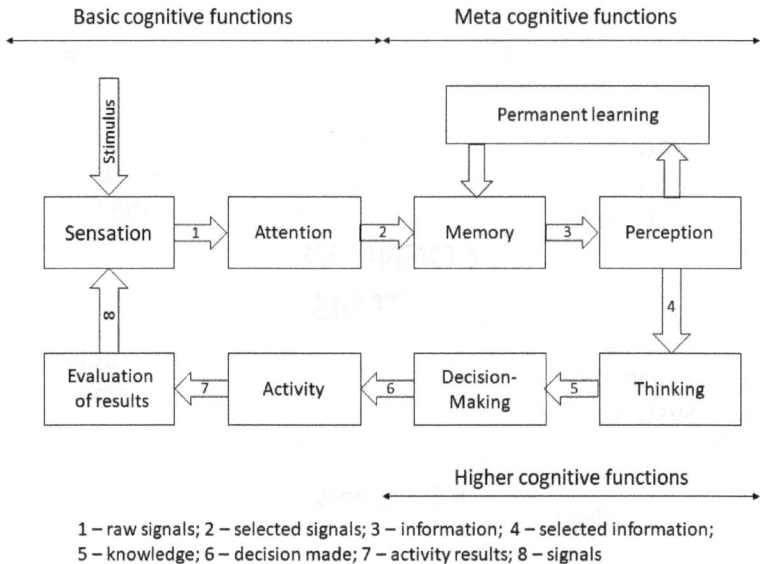

1 – raw signals; 2 – selected signals; 3 – information; 4 – selected information;
5 – knowledge; 6 – decision made; 7 – activity results; 8 – signals

**Figure 5.3:** Cognitive process as a chain consisting of basic, meta, and higher cognitive functions.

resources. Attention is manifested by an attentional bottleneck, in term of the amount of data the brain can process per unit of time (Ciccarelli and White, 2015).

The third cognitive function is *memory*, the ability to retain information or a representation of past experience based on the mental processes of learning or encoding, retention across some interval of time, and retrieval or reactivation of the memory. Memory is a complex function in which it is distinguished:

- sensory memory (limited capacity and duration to 1–4 sec),
- short-term and working memory (capacity limited to 3–5 items and duration to 12–30 sec),
- long-term memory (capacity unlimited, duration permanent):
  - associated with physical changes in the brain;
  - different types, based on information stored (implicit memory for skills, habits, and learned responses, and explicit memory for facts and semantic or episodic information);
  - organized in terms of related meanings and concepts.

*Perception* is the organization, identification, and interpretation of sensation  to represent and understand the presented information. Perception is not only the passive receipt of these signals, but it is also shaped by the learning, memory, expectation, as well as attention. The process that follows connects a person's expectations and attention with his or her knowledge that influence perception. *Learning* is any relatively permanent change in behaviour brought about by experience or practice. Memory plays a central role in the learning process, because without the ability to recall the past, people would not be able to enrich their skills and deepen their knowledge. Learning to make voluntary responses through the effects of positive or negative consequences is called *operant conditioning*.

*Thinking* refers to mental activities that occur in the brain when processing, organizing, understanding, or communicating information to others. It is strongly connected with so-called mental imagery, which creates representations for objects or events used in mental activities. Mental images are interacted with in similar ways as physical objects and processed in the brain slightly differently than actual objects. *Concepts* are ideas that represent a class or category of objects, events, or activities used to interact and organize information without having to think about or process every specific example of the category. Concepts represent different levels of objects or events, can be well-defined based on strict criteria, or fuzzy, based on personal experience are represented by prototypes.

The last cognitive function in the chain is *decision-making*. It is the cognitive process of choosing between two or more alternatives, ranging from the relatively clear cut to the complex. Psychologists have adopted two converging strategies to understand decision-making (https://dictionary.apa.org/decision-making):

- statistical analysis of multiple decisions involving complex tasks, and
- experimental manipulation of simple decisions, looking at the elements that recur within these decisions.

Both these strategies build on the knowledge provided to the 'Decision-Making' block from the previous 'Thinking' block. The other two elements of the chain, namely Activity and Evaluation of results, do not belong to the cognitive functions and will therefore not be discussed in detail here.

## 5.3 Processes of Knowledge Acquiring in Engineered Systems

Acquisition of knowledge necessary to develop and manage engineering systems is based on the cognitive processes. Aristotle described these processes in Metaphysics (Irwin, 1988) and identified three main interrelated levels: experience, knowledge, and wisdom. Based on this concept, in the second half of the 20th century a model of 'The pyramid of knowledge hierarchy' was created, in which the level of 'experience' was divided into two elements: data and information, and the other two levels – knowledge and wisdom – remained unchanged (Ackoff, 1989). This model has become the basis of a dynamically developing concept of the knowledge management and has undergone several modifications.

The pyramid of knowledge is based on data which are described by symbols and represent individual observations of real-world states. *Data* represent raw facts, events, or statements without reference to other things. It does not have a meaning of itself and has no inherent structure. Data can be measurable or not, analogue, or discrete, as well as considered statically (e.g., data record) or dynamically (e.g., data stream). It can be obtained from different sources.

*Information* is data that are processed to be practical useful, it means data that has been given meaning in relation to other data, relevance for a user and purpose. Any information is partially subjective because it depends not only on the data, but also on the process of their interpretation, which is based on the knowledge held by the interpreter at that time. Thus, information can be defined as a collection of selected data, processed, and presented in a form that should be useful to the decision-maker. The main condition for the usefulness of information is its ability to be interpreted in a specific context, and to find answers to simple questions like: Who? What? Where? and When? These dependencies can be explained by the Langenfor's model, which represents the following relation, called the infological equation (Skyttner, 2008):

$$I = i(D, K, t) \tag{5.1}$$

where:  $I$ – information achieved by an interpretation process,

$I$ – an interpretation process acting on data,

$D$ – available data,

$K$ – previous knowledge,

$t$ – available time.

It follows from model (5.1) that information is a certain abstract relationship, and thus without context it cannot be objectively interpreted. Information can be structured according to the following: alphabet order, content category, continuum (continuous or discontinuous), location and time.

*Knowledge* will be understood as the sum of facts and principles gained through experience, learning, and thinking, leading to an understanding of complex issues (e.g., phenomena, processes, systems). It can be created by integrating new information with existing knowledge about a particular area of interest. Knowledge acquisition requires the ability to evaluate available information and understand reality in the light of this information, in accordance with the current state of knowledge. In practice, the use of knowledge is often reduced to looking for answers to complex questions, such as 'how?'. Knowledge is emergent in relation to information, e.g., by applying processes of systematization and structuring of information.

*Wisdom* is understood as the wide and deep knowledge combined with intelligence and maturity. It is characterized by the following features:

- ability to make legitimate decisions that bring positive results in the future,
- ability to make effective use of your knowledge and skills,
- ability to understand the world, phenomena, and relationships between them.

Thus, the notion of wisdom places itself at the edge of many scientific disciplines, and above all, philosophy.

The interdependences between these four categories: data, information, knowledge, and wisdom, are shown in Figure 5.4a the traditional form, and in Figure 5.4b the modified knowledge pyramid for engineered

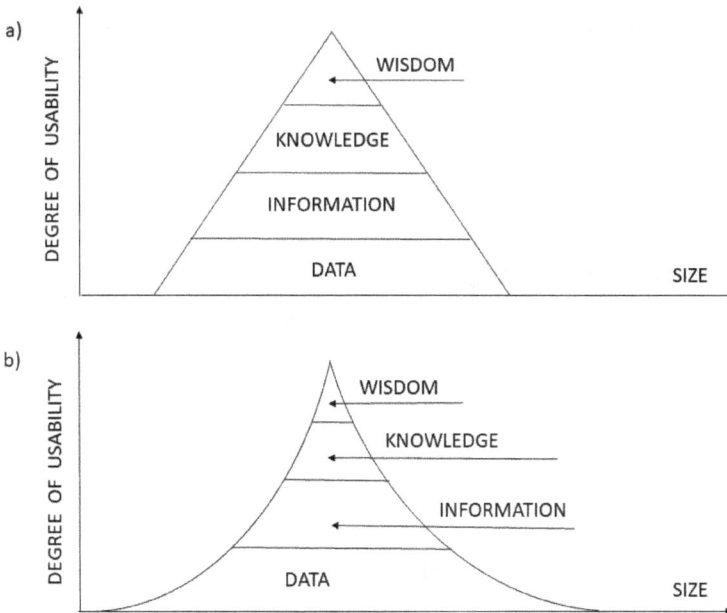

**Figure 5.4:** The knowledge pyramid: (a) traditional approach, (b) modified approach for engineered systems (based on Bukowski, 2019).

systems. The proposed modification allows for a quantitative assessment of the usability degree for each category and, to consider the quantity of units in each category (the number of units in each category is proportional to the area representing this category). This model has a static nature and can be interpreted as a hierarchical metaphor of the relationship between the various elements of the cognitive process. Considering the specificity of cognitive processes shown in Figure 5.3, a simplified model of knowledge acquisition process in engineered systems can be described in terms of the diagram shown in Figure 5.5. The starting point of the process is the row data acquisition, from which the data stream (1) is sent to the data processing block. The selected data is then sent (2) to a memory block, from which it is retrieved by another block to form the information (3). In the next block (4), this information is processed to discover relationships between them and to build recurrent patterns (5), which in turn are used to create new knowledge (6). This new knowledge complements the existing knowledge base (7), so that the procedures for acquiring new data can be adjusted (8) and the process of permanent acquisition of knowledge can be continued more effectively.

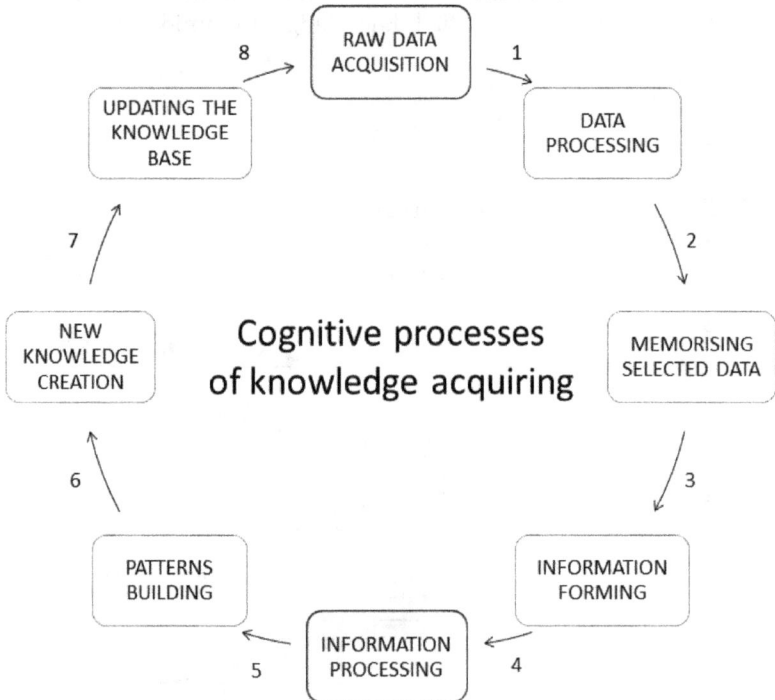

**Figure 5.5:** A simplified model of knowledge acquisition process in engineered systems.

## 5.4  Defining Cognitive Systems Engineering – Imperfect Knowledge Acquisition and Management Concept

The concept of *imperfect knowledge* (Krause and Clark, 1993) is taken as a starting point when considering how to describe Cognitive Systems Engineering. In 1957, G. Bergmann published his work, *Philosophy of Science* (Bergmann, 1957), in which he made a distinction between 'perfect' and 'imperfect' knowledge. *Perfect knowledge* is based on complete theories and deterministic nature laws. It concerns closed systems in which the initial conditions are known accurately, and the only method used in the inference is the deductive method. Whereas, in applied sciences such as systems engineering, social science, economics or management, knowledge is based on incomplete theories as well as indeterministic rules and principles. It concerns open systems in which the initial conditions are known inaccurately, and the dominant method used in the inference is the inductive method, which generates imperfect results. The data comes from observations or measurements whose accuracy is limited and therefore is uncertain. If it is possible to repeat the same observations or measurements many times, the use of statistical methods minimizes the impact of these imperfections on our knowledge. But in many cases, there is no such possibility, and then we must settle for knowledge withe some level of imperfection.

Figure 5.6 shows the knowledge extraction process as a chain of actions that must be performed to obtain knowledge from raw data that is useful for making complex decisions.

These activities can be divided into five main steps, which will be discussed below.

I.  Data processing

Data imperfection, for whatever reason, can result from its:

- imprecision (inaccuracy) – data subject to measurement error of a random, systematic nature or due to insufficient resolution,
- vagueness – data that is ambiguous, improperly defined,
- incompleteness – data with significant missing elements,
- incoherency – data that are inconsistent or contradictory,
- implausibility – incredible, unimaginable, inconsistent with reality.

To eliminate these imperfections, or at least reduce them to a minimum, the raw data, which may appear in structured, unstructured, sensory, or other forms, should be processed. In the data processing step, four main tasks may be differentiated: data cleaning, data integration, data transformation, and data reduction. Below follows a description of

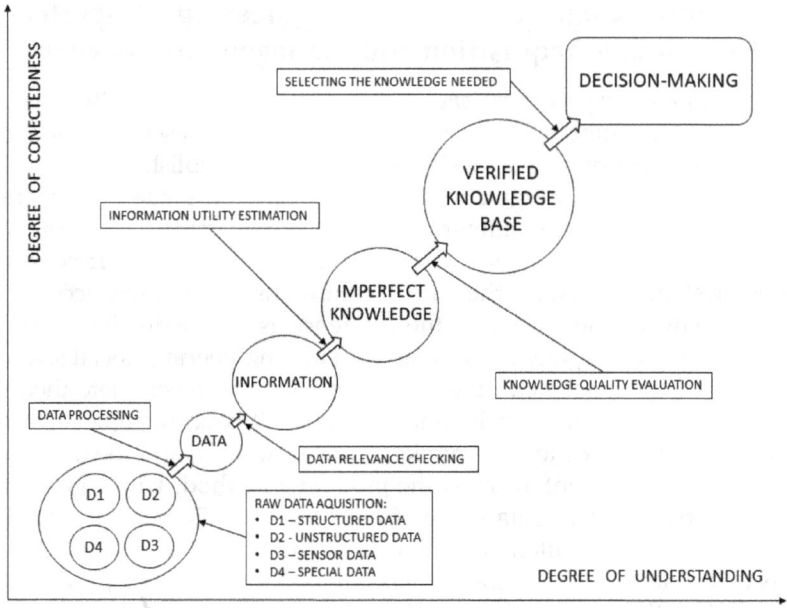

**Figure 5.6:** The knowledge extraction process as a chain of actions that must be performed to obtain knowledge from raw data that is useful for making complex decisions (based on Bukowski, 2019a).

the typical procedures involved in these tasks (based on Al Shalabi et al., 2006):

a) Data cleaning

The data cleaning task involves three operations, which can be supported by methods supported by appropriate computational algorithms:

- replacement of missing or empty values by calculated numbers with the use of the remaining attribute values,

- accuracy improvement trough the replacement of the current value with the newly calculated one or through the removal of this current value,

- inconsistency removal trough special procedures or codes programmed in data collection sheets.

b) Data integration

For knowledge extraction it is necessary to represent the data in the form of a two-dimensional table because the calculation sheets are the most useful. The operations are performed with the following procedures:

- recognizing the attributes which could not have been identified at the cleaning process,

- redundancy removal by comparing the degree of similarity of the data,
- unification of the data collected in different sets with the same form and the same units.

c) Data transformation

Data transformation encompasses all the issues connected with converting the data into a form which makes data exploration convenient, and involves the following six operations:

- smoothing by elimination of the local data deviations having the character of noise (e.g., using the techniques, such as binning, clustering, or regression),
- aggregation by summing up the data (e.g., as the function of time),
- generalization by converting the collected data into higher-order quantities (e.g., via their discretization),
- normalization by the rescaling or adjustment of the data to a specified, narrow range (e.g., from 0 to 1),
- attribute construction by mathematical transformations with the aim of obtaining a new feature, which will replace its constituent attributes in modelling,
- accommodation by transforming the data into a format used by specific computational algorithms.

II. Data relevance checking

Uncertainty of information can have objective causes, resulting from imperfections in the data, and subjective causes, caused by errors in the process of interpretation. In practice, objective causes generally dominate, so information uncertainty is mainly determined by the imperfection of the data held. It is proposed to adopt the concept of data relevance as a measure of the usefulness of individual single data to create reliable information from them.

*Data relevance* (DR) is the property of the data described by five-dimensional vector of its attributes:

- Accuracy (ACC) – the degree to which the data fulfils the relevant requirements (the data should precisely meet certain specifications and standards).
- Clarity (CLA) – the degree to which the data can be clearly understood (the data should be well defined, without several meanings, vagueness, and ambiguity).

- Consistency (CON) – the degree to which the data is compatible with the same type of data from different sources (the data should be coherent without confusing or conflicting meaning).
- Plausibility (PLA) – the degree to which the data is fitted to reality (the data should be compatible with reality, imaginable and possible).
- Traceability (TRA) – the degree to which the data can be traced to its sources (the origin of the data should be ascertained with confidence).

This definition can be illustrated by the model, which represents the following relation:

$$DR = (ACC, CLA, CON, PLA, TRA) \qquad (5.2)$$

III. Information utility estimation

The degree of information imperfection determines its quality and usefulness for creating new knowledge. The most common dimensions of *information quality* are following (Bukowski, 2019):

- Accessibility – extent to which information is available, or easily and quickly retrievable.
- Accuracy – extent to which data are correct, reliable, and free of error.
- Amount of data – extent to which the quantity or volume of available data is appropriate.
- Availability – extent to which information is physically accessible.
- Believability – extent to which information is regarded as true and credible.
- Completeness – extent to which information is not missing and is of sufficient breadth and depth for the task at hand.
- Concise – extent to which information is compactly represented without being overwhelming (i.e., brief in presentation, yet complete and to the point).
- Consistency – extent to which information is presented in the same format and compatible with previous data.
- Efficiency – extent to which data can quickly meet the information needs for the task at hand.
- Navigation – extent to which data are easily found and linked to.
- Objectivity – extent to which information is unbiased, unprejudiced, and impartial.
- Relevancy – extent to which information is applicable and helpful for the task at hand.

- Reliability – extent to which information is correct and reliable.
- Reputation – extent to which information is highly regarded in terms of source or content.
- Security – extent to which access to information is restricted appropriately to maintain its security.
- Timeliness – extent to which the information is sufficiently up to date for the task at hand.
- Understandability – extent to which data are clear without ambiguity and easily comprehended.
- Usability – extent to which information is clear and easily used.
- Usefulness – extent to which information is applicable and helpful for the task at hand.
- Value-Added – extent to which information is beneficial and provides advantages from its use.

In practice, the quality of information is conditioned by the context in which it is used. Therefore, in the decision-making process it is essential to evaluate the usefulness of information as its quality in the context of the specific purpose of its use. Based on the literature (Kulikowski, 2014) it is proposed the introduction of the term 'information utility' as an equivalent of the usefulness of information, and the following definition of the term.

*Information Utility* (IU) is the property of the information described by five-dimensional vector of its attributes:

- Believability (BEL) – the degree to which the information can be reliable (the information should be believable, of undoubtful credibility, and from a reputable source).
- Completeness (COM) – the degree to which the information does not contain omission errors (the information should: include all the necessary values, be complete, cover the needs of our tasks and have sufficient breadth and depth).
- Correctness (COR) – the degree to which the information is proper (the information should be free from errors).
- Relevancy (REL) – the degree to which the information is useful (the information should be relevant and applicable to our work, as well as appropriate for our needs).
- Timeliness (TIM) – the degree to which the information is up to date (the information should be sufficiently: timely, current for our work and fresh enough for our needs).

This definition can be illustrated by the model, which represents the following relation

$$IU = (BEL, COM, COR, REL, TIM) \tag{5.3}$$

IV. Knowledge quality evaluation

Information that has been positively verified as useful forms the basis for creating new knowledge. This process can be described using cognitive models based on the following assumptions (Andrzejczak and Bukowski, 2021; Wang, 2015, 2016):

1) Data, information, and knowledge are cognitive objects obtained because of abstraction.

2) Abstraction is a cognitive process of eliciting a target set of attributes of specific objects based on available data or information.

3) Elicitation is the process of objectively synthesising subjective evaluations of cognitive objects under the constraints of uncertainty inherent in the problem under consideration.

4) Knowledge can be described by the triplet

$$K = (O, A, R) \tag{5.4}$$

where:

$K$ – set of conceptual knowledge,
$O$ – set of objects,
$A$ – set of attributes characterizing $O$ objects,
$R$ – set of relations between objects $O$ and attributes $A$.

whereby:

$$R = (O \times A) \tag{5.5}$$

5) New knowledge $K_N$ is created by mapping useful information $I$ onto a complementary and non-contradictory set of formal concepts $C$ which can be written as follows:

$$K_N = f_k: I \rightarrow C \tag{5.6}$$

Starting from the above assumptions, the following knowledge evaluation quality $KQ$ model is proposed:

$$KQ = (DR, IU, CP) \tag{5.7}$$

where:

$DR$ – data relevance,
$IU$ – information utility,
$CP$ – concepts plausibility understood as the degree to which the data is fitted to reality (e.g., imaginable, and possible).

In the next step, the process of verification of imperfect knowledge in terms of the quality of its evaluation takes place and only that part of it which passes this process is placed in the database of verified knowledge. When making specific decisions, only a selected part of knowledge collected in the database is used each time.

The approach to cognitive systems and processes described above, based on the concept of imperfect knowledge, provides the basis for defining a new field of knowledge:

COGNITIVE SYSTEMS ENGINEERING, HAS EMERGED AS A RESULT OF SYNERGETIC INTERACTION BETWEEN THREE DISCIPLINES OF APPLIED SCIENCES, NAMELY: SYSTEMS ENGINEERING, KNOWLEDGE ENGINEERING, AND COGNITIVE COMPUTING, AND ITS OBJECTS OF APPLICATION ARE PHYSICAL SYSTEMS, CYBER SYSTEMS, AND SOCIAL SYSTEMS, AS WELL AS ANY COMBINATION THEREOF (E.G., CYBER-PHYSICAL-SOCIAL SYSTEMS).

A simplified diagram of these relationships is shown in Figure 5.7.

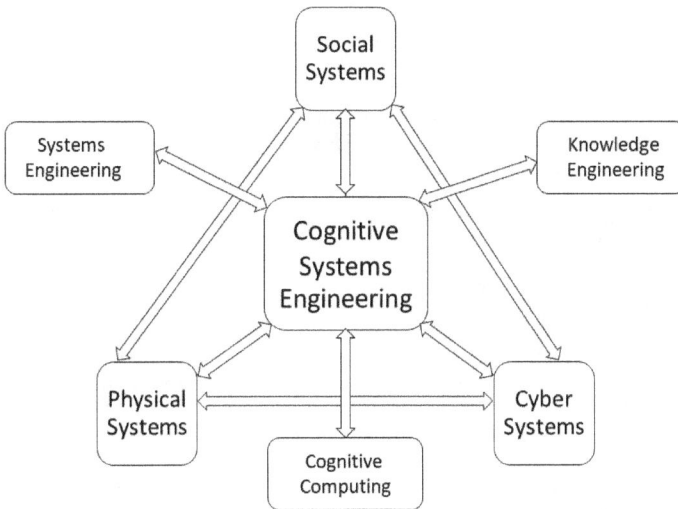

**Figure 5.7:** Cognitive Systems Engineering and its relationship to other areas of knowledge.

Presented in Section 2.3 the concept of the Viable System (Figure 2.2 shows its general model) seems to be the most appropriate to create a structural model of Cognitive Engineered Systems. A proposal for such a model is shown in Figure 5.8. It consists of a Physical System and a Cyber System whose interaction is coordinated by metasystem Cognitive Governance Metasystem (CGM). The Physical System consists of n subsystems that are integrated and coordinated through metasystem CGM, while the Cyber System is composed of subsystems: Sensing and

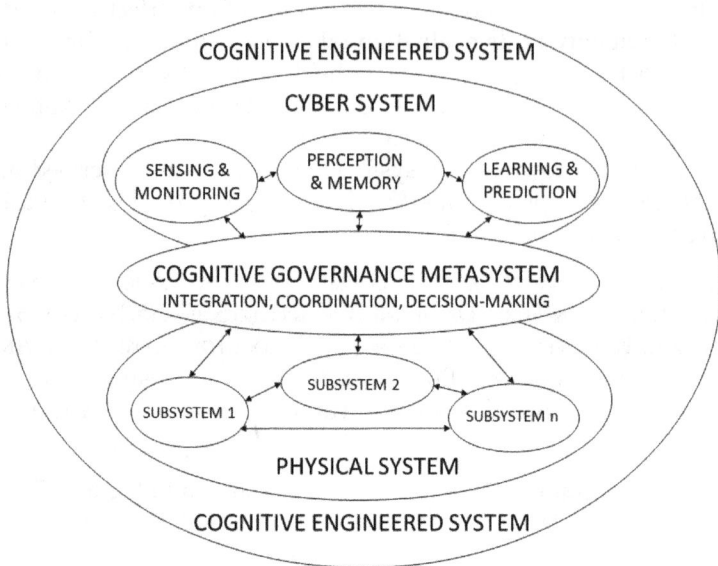

**Figure 5.8:** A structural model of Cognitive Engineered Systems.

Monitoring, Perception and Memory, as well as Learning and Prediction. Based on the information provided by the Cyber-System, CGM makes decisions, in terms of management as well as governance of the entire Cognitive Engineered System. This model will be used in the dependability analysis of cognitive systems in Chapter 11.

## 5.5 References

Ackoff, R.L. (1989). From data to wisdom. *Journal of Applies Systems Analysis*, 16: 3–9.

Al Shalabi, L., Shaaban, Z. and Kasasbeh, B. (2006). Data mining: A preprocessing engine. *Journal of Computer Science*, 2(9): 735–739.

Andrzejczak, K. and Bukowski, L. (2021). A method for estimating the probability distribution of the lifetime for new technical equipment based on expert judgement. *Eksploatacja I Niezawodność – Maintenance and Reliability*, 23(4): 757–769. http://doi.org/ 10.17531/ ein.2021.4.18.

Bergmann, G. (1957). *Philosophy of Science*. University of Wisconsin Press.

Bukowski, L. (2019). *Reliable, Secure and Resilient Logistics Networks. Delivering Products in a Risky Environment*, Springer Nature Switzerland AG 2019. ISBN 978-3-030-00849-9 (Hardcover), ISBN 978-3-030-00850-5 (eBook).

Bukowski, L. (2019a). Logistics decision-making based on the maturity assessment of imperfect knowledge. *In: Engineering Management in Production and Services*. ISSN: 2543-6597 (print), 2543-912X (online), 11(4): 65– 79; Doi: 10.2478/emj-2019-0034.

Ciccarelli, S.K. and White J.N. (2015). *Psychology*. 4th Edn. UK: Pearson.

https://dictionary.apa.org/cognitive-process.

https://dictionary.apa.org/decision-making.

https://plato.stanford.edu/entries/cognitive-science/.

Irwin, T. (1988). *Aristotle's First Principles.* Oxford: Clarendon Press.

Krause, P. and Clark, D. (1993). *Representing Uncertain Knowledge. An AI Approach.* Oxford: Intellect.

Kulikowski, J.L. (2014). Data quality assessment: Problems and methods. *International Journal of Organizational and Collective Intelligence*, 4(1): 24–36.

Lakoff, G. and Johnson, M. (1999). *Philosophy in the Flesh: The Embodied Mind and Its Challenges to Western Thought.* Chicago: University of Chicago Press.

Skyttner, L. (2008). *General Systems Theory Problems, Perspectives, Practice.* Singapore: Word Scientific.

Talmy, L. (2000). *Toward a Cognitive Semantics.* Cambridge, Mass.: The MIT Press.

Wang, Y. (2015). Formal cognitive models of data, information, knowledge, and intelligence. *WSEAS Transactions on Computers*, 14: 770–781.

Wang, Y. (2016). On cognitive foundation and mathematical theories of knowledge science. *International Journal of Cognitive Informatics and Natural Intelligence*, 10(2): 1–19. https://www.researchgate.net/publication/303810684.

Wolfson, L.J. (2001). *International Encyclopedia of the Social and Behavioral Sciences.* https://www.sciencedirect.com/topics/computer-science/elicitation-process.

Part II

# Concepts of Attribute-oriented Systems Engineering

From Reliability Engineering to
Cognitive Dependability Engineering

# 6

# Reliability Engineering
## The Concept of
## Failure-free Operation*

This chapter provides an overview of the major concepts that underpinned the emergence and dynamic development of Reliability Engineering as an interdisciplinary field of knowledge. Physical aspects of reliability are also discussed, as being particularly relevant to the engineering approach, and neglected in most works on Reliability Engineering. The mathematical basis for the evaluation of the reliability of elements and entire systems is presented, both based on mathematical statistics and on the theory of probability. Models of selected discrete probability distributions as well as continuous probability distributions are presented, and the calculation formulas necessary to determine the parameters of these distributions are summarized. Selected examples of metrics for reliability attributes have been proposed and illustrated with examples. The section on structural reliability discusses the Reliability Block Diagram method and its application to assess the reliability of systems with different functional structures. The computational models for selected redundant structures are presented, namely so-called $(k, n)$ systems as well as with hot, warm, and cold reserve. The conclusion presents the limitations of the application of these theoretical models in real operating conditions.

* https://orcid.org/0000-0002-2630-3507.

## 6.1 The Emergence and Development of Reliability Engineering

Generally, reliability can be considered as a time-oriented quality (O'Connor and Kleyner, 2012). Thus, it refers to the future performance or behaviour of a component or system. More precisely, reliability is the ability of a product or system to perform as intended (i.e., without failure and within specified performance limits) for a given time, in its lifecycle conditions (Kapur and Pecht, 2014). *Reliability Engineering* is a sophisticated interdisciplinary field which requires a wide-ranging body of knowledge in the basic sciences, including physics, chemistry, and mathematics, an understanding of the broader issues within system integration and engineering, as well as considering costs and schedules.

Reliability Engineering as a scientific discipline was established in the mid-1950s. The formation of the Advisory Group on the Reliability of Electronic Equipment (AGREE) by the US Department of Defense in 1956 can be considered as the turning point in reliability development (Sage and Rouse, 1999). Most of the methodologies available for reliability assessment were originally developed for electrical systems; however, it is easier to perform repetitive tests to produce many failure samples in a short time for electronic systems. This was the main reason for the development of the reliability definition on a statistical basis. Because the hazard rate of typical electronic items can be well approximated by the constant failure rate model, the exponential distribution became systematically and widely used as the reliability model. The main driving force for its popularity was the simplicity of the corresponding reliability functions. This made the exponential distribution the dominant model in early reliability assessments.

One decade later, in the 1960s, the exponential distribution turned out to be not so practical for many applications and sensitive to departure from the initial assumptions. The application of this model for components with high reliability targets could result in unrealistic results of mean-time-to-failure (MTTF) estimates (Sage and Rouse, 1999). Additionally, this model ignored any aging and degradation in the component. Similarly, a repairable item was expected to experience the exact same rate of occurrence of failure as the new one, which is also unrealistic. After such disappointments, reliability practitioners tried to capture some of the physical characteristics of failure into their modelling by using other available traditional distributions, such as the Weibull and lognormal distributions. The hazard rate for the Weibull distribution is time-dependent and can be either monotonically decreasing or increasing, which explains infant mortality and aging. In addition, no repair activities are usually required for electronic components within a system, since failed components are simply replaced by new parts.

In contrast to electronic systems, there are typically no abrupt failures in mechanical systems, and degradation processes leading to failure happen usually gradually (e.g., frictional wear or fatigue cracking). This means that the assumption of memory-less exponential distribution for electronic products, it is not the case for mechanical systems due to the size, durability, and maintainability of such systems. Thus, for these systems other life models with variable hazard rate appeared to be a better option. The concept of *physics of failure* (PoF) was introduced in the early 1960s as part of a series of symposia that took place at the Rome Air Development Center (Azarkhail and Modarres, 2012).

The 1970s marked the birth of *fault tree analysis* (FTA), which was motivated by the need for safety assessment in the aerospace industry and later for nuclear power plants (Keller and Modarres, 2005). During those years, most reliability engineering efforts were focused on reliability of components and devices. There was intense interest in safety, risk, and reliability in different applications such as the gas, oil, and chemical industries, and especially, in nuclear power applications. The electronic system origins of reliability assessment can also be seen in the appearance of parallel and series configurations in reliability block diagram as well as fault tree analysis and event tree applications. The operation of electronic components in parallel or series has a small and sometimes negligible impact on each other that made the independent-event assumption a very common practice in early reliability modeling attempts.

In mechanical systems, because of high costs of such redundancies, the design concept of the weakest link appeared often enough to define the failure logic of the system. Also, mechanical components usually operate in highly varied dynamic environments in which the operational condition of one component strongly depends on the neighbouring ones. The functional dependency of mechanical components acts through operational conditions such as temperature, pressure, lubrication, and other transient characteristics of the system dynamic, all of which should be addressed in the reliability model of the components. The community of mechanical engineers proposed some early parametric, data-driven methods to address presence of functional dependencies at the system-level analysis.

The 1980s experienced explosive growth in integrated circuit (IC) technology. The traditional approach for such components was to collect as much field failure data as possible to build a statistical model for the component life. For ICs, however, the collected data showed a strong correlation between the failure rate and the complexity of the ICs. This complexity, measured by the number of gates and transistors, was successfully incorporated into the life model of ICs (Denson, 1998). Measures such as defect density, the die area and the yield of the die were introduced later as different physical measures to be considered along with the statistical life models. Due to faster trends in mass production, great emphasis was placed on capturing the needed information with much less effort. As a result, design and assessment methodologies that addressed the *root causes of failure* and other operating conditions emerged as efficient cost-saving techniques. The *accelerated life modelling* approach was a result of this movement. This model could have included some of the operational conditions and made the life models more flexible. Such models were more universal, and needed much less failure data (Nelson, 1990). Many assumptions and simplifications need to be made, which could unacceptably limit the relevance of the outcomes. In addition, the accelerated life models, like all other statistical-based approaches, needed data for validation, which was not available for a product in the design stage or for a highly reliable product. In such cases, reliability models could be constructed based on some generic data from the history of similar products. This data could be updated later with expert judgements (elicitations) or other data generated with statistical inference techniques. Therefore, more dependable reliability techniques needed to be developed to meet the requirements of mass manufacturing technologies.

The other trend in the 1980s related to the growing application of Bayesian method in probabilistic data analysis. Using this approach, engineers utilized data from expert opinions and any previous experience with similar products to make a probability density referred to so-called prior distribution. The Bayesian framework made it possible to update this imperfect knowledge later and by a few new data to make an upgraded posterior state of knowledge (Azarkhail and Modarres, 2007). However, the integrals necessary for normalization at Bayesian conditional probability calculations can be very complex when dealing with multi-parameter reliability models, also the applications of this approach were originally limited to simple reliability models.

As noted earlier, the dependency of failures can be a critical factor in reliability modelling of mechanical systems and components. There are situations in which progressive failure of one component may activate or accelerate the failure of others, or one failure mechanism may impact other mechanisms of the components' degradation. In practice, many links have been found between different components by means of their properties and environmental conditions. The 1980s also marked the development of initiatives for modelling dependencies at the system level with the common-cause failures approach. The *common-cause failure* (CCF), which is the failure of more than one component due to a shared root cause, is classified as a dependent failure (Fleming and Mosleh, 1985).

In the 1990s, there was a resurgence of interest in physical methods (PoF) in reliability studies. This was mainly due to the use of ever faster computers and more powerful software, especially in fracture mechanics, the foundations of which were developed by Alan A. Griffith and later George R. Irwin. In this approach, facts from root-cause physical and chemical degradation processes are used to prevent the failure of the products by robust design and better manufacturing practices (Pecht and Dasgupta, 1995). As a result, both design and assessment methodologies that address the root causes of failure have emerged as powerful and cost-saving techniques. The US Army and Air Force initiated two reliability-physics related programmes, e.g., to promote a scientific approach to reliability assessment of electronic equipment (Ebel, 1998). This approach to reliability allowed for the identification of potential damage and failure mechanisms that individually or, in combination, lead to the item's failure. In addition, properly developed and tested PoF models, would be used to estimate expected durability beyond the time scale of the experimental tests. Such reliability models allow to describe quantitatively the process of degradation of elements of technical systems, giving practical answers to questions of the type "how and why the item fails" and additionally reduce the need for large quantities of life data.

In the late 1990s, criticism against the use of catalog data collected in documents such as MIL-HDBK-217 became increasingly intense, and the basic idea of using failure rate data gathered in such databases was increasingly questioned. However, the critics had difficulties proving that the PoF approach produces more accurate reliability estimators (Fragola, 1996). Most of the PoF models strongly depended on experimental data, so the basic question is: if there is enough data available to evaluate a PoF-based model, why not use the same data for determining failure rate by statistical methods, or even turn to an integrated use of both approaches? In practice, the uncertainty of estimates is often so large that they become useless in the process of making operational decisions. To master the problem of uncertainty, two main factors needed to be considered. First, get more data for which accelerated life testing, step-stress testing, expert judgment, and many different resources were exhausted. The second factor was an appropriate computational frameworks and software which allow new data to be easily added to the analysis. One of the possible choices is the *classical maximum likelihood estimation method*, based on an implied Bayesian uniform prior for the parameters, leading to the most probable set of values for the parameters (Hald, 1999). This method used the likelihood function of the available data for the model parameter estimation and provided no means to incorporate prior knowledge available for the model parameters. However, this method fails if no complete failure data was available which is the case for modern, highly reliable components and systems, and even if some failures become available in the lab, it is usually difficult to associate them with the field-observed failure mechanisms. Additionally, when using the maximum likelihood estimation approach the mean effect of the data is often masked due to over-reliance on the mode of the likelihood function, especially when dealing with small sample sizes.

In contrast with classical maximum likelihood estimation method, the Bayesian approach (Congdon, 2003) provided many useful features, among other things, providing opportunities to use a priori knowledge and the possibility of using many different forms of data (e.g., directly from measurements, censored, fuzzy, partially relevant, and expert judgements). However, one of the most important limiting factors of the Bayesian inference methods in practical reliability analysis was the mathematical complexity of the problem, due to multidimensionality. In the 2000s, the numerical and computational advancements in Bayesian statistical methods, such as Markov Chain Monte Carlo simulations (Brooks, 1998), joint with advancements in computational tools, made the Bayesian inference techniques increasingly useful in practice. Reliability practitioners combined this approach with different types of data (e.g., simple failure rates from handbooks, engineering expert judgements, simulated results of PoF models, and direct test results) in a sophisticated hybrid platform. Supported by fast computing, hybrid methodology became widely available and practical useful. These techniques can rely on the physical and chemical phenomena that drive degradation and failures, combined with accelerated tests and field or expert judgment data.

To summarize this brief historical overview, it can be concluded that Reliability Engineering is a sophisticated and demanding interdisciplinary field. Today, there are basically three characteristics of systems and components that are not appropriately addressed in conventional reliability modelling approaches. These features are dynamic behaviour, complexity, and interdependency of the systems and components. These features have become particularly important for the new generation of complex systems due to the highly competitive market and difficult to predict changes in the future.

## 6.2  Mathematical Foundations of Reliability Engineering

The essential pillars which support the reliability engineering as a scientific discipline are statistics and the theory of probability. The quantitative description of reliability therefore requires the basic tools used in statistics and probability calculus, which will be briefly discussed below (based on Bukowski, 2019).

### 6.2.1 Basics of Statistical Methods

Mathematical statistics include description and analysis of repetitive phenomena using the assumptions of probability theory. Statistical methods allow to draw quantitative conclusions about a general population, understood as a set of elements $\{E\}$ subject to research due to one or more features (e.g., $X$), based on studies of so-called *random sample*. The $n$-element random sample is a finite subset of elements of the general population, drawn in such a way that each subset consisting of $n$ elements of the general population has the same chance of being drawn (Wasserman, 2005).

To conduct statistical analysis, we should take the following steps:

- clearly define the general population $\{E\}$,
- take an n-element random sample from it $\{e_1, e_2, e_3, ..., e_n\}$,
- define the examined feature of the general population $X$,
- determine the values of the $X$ feature for each element of the random sample $\{x_1, x_2, x_3, ..., x_n\}$.

For such a set of numerical values $\{x_1, x_2, x_3, ..., x_n\}$, we determine the basic statistical parameters, which are estimators of the central tendency, dispersion, and frequency distribution of the examined feature $X$ for the whole general population.

The commonly used measures of *central tendency* are:

- arithmetic average
$$x_m = \frac{\sum_1^n x_i}{n} \tag{6.1}$$

- geometric mean
$$x_G = \sqrt[n]{x_1 x_2 ... x_n} \tag{6.2}$$

- median value – when $n$ is an odd number, the central number in an orderly, non-descending sample; and when $n$ is an even number, the arithmetic average of the two middle numbers.

- modal value (dominant) – the most frequently repeated, greater than the smallest and the smallest value.

The commonly used measures of *dispersion* are:

- Range
$$R = x_{max} - x_{min} \tag{6.3}$$

- Interquartile range (IQR) – the difference between the upper and lower quartiles, contains 50% of the total number of the set, where:
  - the $p$-th percentile in the $n$-ordered set of numbers (e.g., random sample) is the value below which $p\%$ of the numbers of this set are located,

- o lower quartile – 25th percentile,
- o upper quartile – 75th percentile.
- Variance (arithmetic mean of squared deviations from the mean value)

$$V = \frac{\sum_1^n (x_i - x_m)^2}{n-1} \qquad (6.4)$$

- Standard deviation
$$s = \sqrt{V} \qquad (6.5)$$

- Coefficient of variation
$$v = \frac{s}{x_m} \qquad (6.6)$$

Typical *frequency distribution* – histogram

In the case of a small random sample size (up to 20), the data is sorted in a non-decreasing series and graphically represented in the coordinate system $(X, N)$, where $X = \{x_1, x_2, x_3, ..., x_n\}$ are elements of the series, while $N = \{n_1, n_2, n_3, ..., n_m\}$ is frequency of individual elements of the series. If the values in the random sample are not repeated, then $m = n$.

In the case of a larger random sample size (over 20), the sample values are grouped in classes, i.e., in intervals (usually the same length), with a size of 5 to 20. The procedure takes place in the following steps:

- The number of classes is taken from an approximate formula

$$k \cong \sqrt{n} \qquad (6.7)$$

- Absolute frequency $n_i$, it is the number of results in each class.
- Relative frequency

$$r_i = \frac{n_i}{n} \qquad (6.8)$$

- Accumulated frequency (absolute or relative) – the sum of the frequency values from the beginning of the distribution series to a given class.
- Distribution series – a series created from pairs of numbers representing the means of successive classes and their numbers.
- Histogram – distribution series presented in graphic form.
- Frequency polygon (broken frequency) – curve obtained by combining the means of successive classes on the graph in the form of a histogram.

## 6.2.2  Basics of Probabilistic Methods

Repeatable exposures (e.g., threats or hazards) may occur in the form of individual events or event chains leading to a failure. These exposures are usually of a random nature and can be modelled in the form of random processes. A *random process* (RP) may be defined as an ensemble of given time functions, any one of which might be observed on any trial of an experiment or real process realization. The ensemble may include a finite number, a countable infinity, or a non-countable infinity of such functions. We will denote the ensemble of functions by $\{X(t)\}$, and any observed member of the ensemble by $x(t)$. The value of the observed member of the ensemble at a particular time $t_i$, is a *random variable*; on repeated trials of the experiment, $x(t_i)$ takes different random values. The probability that $x(t_i)$ takes values in a certain range is given by the *probability distribution function* (PDF), as it is for any random variable. Random process can be continuous or discrete (Joyce, 2016; Mongomery and Runger, 2003; Ross, 2004; Scheaffer et al., 2011; Soong, 2004).

Real random process, also called *stochastic process* (e.g., noise source), can be characterized by its properties, e.g., PDF, *cumulative distribution function* (CDF), mean (expected value), variance, auto-correlation function (statistical average of the product of random variables), and cross-correlation function (measure of correlation between sample function amplitudes of processes $x(t)$ and $y(t)$ at time instants $t_1$ and $t_2$). We can characterize random processes based on how their statistical properties change. If the statistical properties of a random process RP don't change with time, we call the RP *stationary*. The computation of statistical averages (e.g., mean and autocorrelation function) of a random process requires an ensemble of sample functions (e.g., data records) that may not always be feasible. However, to calculate the averages from a single data record is possible only in certain random processes called *ergodic processes*. The ergodic assumption implies that any sample function of the process takes all possible values in time with the same relative frequency that an ensemble will take at any given instant.

We distinguish three main random processes, which form the basis for modelling various phenomena occurring in engineered systems: the Bernoulli process, the Poisson process, and the Gaussian process. Other typical random processes are derived from these three processes as generalizations or specific cases (e.g., Markov process). Below we will briefly discuss these three random processes and the resulting probability distributions of random variables that are applicable to modelling of typical exposures (based on Joyce, 2016).

### 6.2.3 *The Bernoulli Random Process*

A single trial for a *Bernoulli process*, called a Bernoulli trial, ends with one of two outcomes; one called success (e.g., occurrence of the event $E$), the other called failure (e.g., no event $E$ occurrence). Success occurs with probability $p$ while failure occurs with probability $q = 1 - p$. The Bernoulli process consists of repeated independent Bernoulli trials with the same parameter $p$. These trials form a random sample from the Bernoulli population. If we ask how many successes there will be among $n$ Bernoulli trials, then the answer will have a binomial distribution, *Binomial (n, p)*. The Bernoulli distribution, *Bernoulli (p)*, simply says whether one trial is a success. If we want to know how many trials it will be to get the first success, then the answer will have a geometric distribution, *Geometric (p)*. If the question is how many trials there will be to get the $r$-th success, then the answer will have a negative binomial distribution, *Negative Binomial (p, r)*. Given that there are $M$ successes among $N$ trials, if we ask how many of the first $n$ trials are successes, then the answer will have a *Hypergeometric (N, M, n)* distribution. All these distributions are discrete, whereas the comparison of formulas for calculating their distribution models, mean values and variances is shown in Table 6.1.

**Table 6.1:** Models of selected discrete probability distributions and computational formulas for their parameters.

| Distribution | PDF | Mean | Variance |
|---|---|---|---|
| Bernoulli $(p)$ | $f(0) = 1 - p; f(1) = p$ | $p$ | $p(1 - p)$ |
| Binominal $(n, p)$ | $f(x) = \binom{n}{x} p^x (1-p)^{n-x},$ for $x = 0, 1, ..., n$ | $np$ | $np(1 - p)$ |
| Geometric $(p)$ | $f(x) = p(1-p)^{x-1},$ for $x = 1, 2, ...$ | $1/p$ | $(1 - p)/p^2$ |
| Negative binominal $(p, r)$ | $f(x) = \binom{x-1}{r-1} p^r (1-p)^{x-r},$ for $x = r, r+1, r+2 ...$ | $r/p$ | $r(1 - p)/p^2$ |
| Hypergeometric $(N, M, n)$ | $f(x) = \binom{M}{x}\binom{N-M}{n-x} / \binom{N}{n},$ for $x = 0, 1, ..., n$ | $np$ | $\left(\dfrac{N-n}{N-1}\right) npq$ for $q = 1 - p$ |
| Poisson $(\lambda, t)$ | $f(x) = \dfrac{1}{x!}(\lambda t)^x e^{-\lambda t}$ for $x = 0, 1, 2, ...$ | $\lambda t$ | $\lambda t$ |

### 6.2.4 The Poisson Random Process

A *Poisson process* is the continuous version of a Bernoulli process. In the Bernoulli process, time is discrete, and at each time unit there is a certain probability $p$ that event occurs, the same probability at any given time, and the events at one time instant are independent of the events at other time instants. In the Poisson process, time is continuous variable, and there is a certain rate $\lambda$ of events occurring per time unit that is the same for any time interval, and events occur independently of each other. Whereas in a Bernoulli process either no or one event occurs in a unit time interval, in a Poisson process any nonnegative whole number of events can occur in a time unit. As in a Bernoulli process, we can ask various questions about a Poisson process, and the answers will have various distributions. If the problem is how many events occur in an interval of the length $t$, then the answer will have a Poisson distribution, *Poisson* ($\lambda t$). If the question is how long until the first event occurs, then the answer will have an exponential distribution, *Exponential* ($\lambda$). If we ask how long until the $r$-*th* event, then the answer will have a gamma distribution, *Gamma* ($\lambda$, $r$). The comparison of formulas for calculating their distribution models, mean values and variances is shown in Tables 6.1 and 6.2.

### 6.2.5 The Gaussian Random Process

A *Gaussian process* is a stochastic process, such that every finite collection of those random variables has a multivariate normal distribution, i.e., every finite linear combination of them is normally distributed. Its distributions are related to the central limit theorem which says that sample means and sample sums approach normal distributions if the sample size approaches infinity.

The normal distribution, *Normal* ($\mu$, $\sigma^2$), also called the Gaussian distribution, is ubiquitous in probability and statistics. The parameters $\mu$ and $\sigma^2$ are real numbers, $\sigma^2$ being positive with positive square root $\sigma$ called standard deviation. The standard normal distribution has $\mu = 0$ and $\sigma^2 = 1$. Normal distributions are used in statistics to make inferences about the population mean when the sample size n is large, and in Bayesian statistics as conjugate priors for the family of normal distributions with a known variance.

$\chi^2$–*distribution*, *ChiSquared* ($v$), is continuous. The parameter $v$, the number of 'degrees of freedom', is a positive integer. This is the distribution for the sum of the squares of $v$ independent standard normal distributions. The $\chi^2$–distribution is a special case of a gamma distribution

**Table 6.2:** Models of selected continuous probability distributions and computational formulas for their parameters.

| Distribution | PDF | Mean | Variance |
|---|---|---|---|
| Normal $(\mu, \sigma^2)$ | $f(x) = \dfrac{1}{\sigma\sqrt{2\pi}} exp\left(-\dfrac{(x-\mu)^2}{2\sigma^2}\right)$ <br> for $x \in \mathbb{R}$ | $\mu$ | $\sigma^2$ |
| Log-normal $(\mu, \sigma)$ | $f(x) = \dfrac{1}{x\sigma\sqrt{2\pi}} exp\left(-\dfrac{(\ln(x)-\mu)^2}{2\sigma^2}\right)$ <br> for $x \in [0, \infty)$ | $e^{\mu+\sigma^2/2}$ | $(e^{\sigma^2}-1)e^{2\mu+\sigma^2}$ |
| Chi Squared $(v)$ | $f(x) = \dfrac{x^{\frac{v}{2}-1}e^{x/2}}{2^{\frac{v}{2}}\Gamma(\frac{v}{2})}$, for $x \geq 0$ | $v$ | $2v$ |
| Uniform $(n)$ discrete | $f(x) = 1/n$, for $x = 1, 2, \dots, n$ | $(n+1)/2$ | $(n^2-1)/12$ |
| Uniform $(a, b)$ continuous | $f(x) = 1/(b-a)$, for $x \in [a, b]$ | $(a+b)/2$ | $(b-a)^2/12$ |
| Triangular $(a, b, c)$ | $f(x) = \dfrac{2(x-a)}{(b-a)(c-a)}$ for $a \leq x \leq c$ <br> $f(x) = \dfrac{2(b-x)}{(b-a)(b-c)}$ for $c \leq x \leq b$ | $\dfrac{a+b+c}{3}$ | $\dfrac{a^2+b^2+c^2-ab-ac-bc}{18}$ |
| Weibull $(\alpha, \beta)$ | $f(x) = \dfrac{\beta}{\alpha}\left(\dfrac{x}{\alpha}\right)^{\beta-1} e^{-\left(\frac{x}{\alpha}\right)^\beta}$ <br> for $x > 0; \alpha > 0; \beta > 0$ | $\alpha\Gamma\left(1+\dfrac{1}{\beta}\right)$ | $\alpha^2\Gamma\left(1+\dfrac{2}{\beta}\right)-\alpha^2\Gamma^2\left(1+\dfrac{1}{\beta}\right)$ |
| Exponential $(\lambda)$ | $f(x) = \lambda e^{-\lambda x}$ for $x \in [0, \infty)$ | $1/\lambda$ | $1/\lambda^2$ |
| Gamma $(\lambda, r)$ | $f(x) = \dfrac{1}{\Gamma(r)}\lambda^r x^{r-1}e^{-\lambda x}$ for $x \in [0, \infty)$ | $r/\lambda$ | $r/\lambda^2$ |

with a fractional value for $r$ (*ChiSquared* $(v)$ = *Gamma* $(\lambda, r)$ where $\lambda = 1/2$ and $r = v/2$). $\chi^2$–distributions are used in statistics to make inferences on the population variance when the population is assumed to be normally distributed.

*Student's T–distribution*, $T(v)$, is continuous. If $Y$ is *Normal* $(0,1)$ and $Z$ is *ChiSquared* $(v)$ and independent of $Y$, then $X = Y/(Z/v)^{1/2}$ has a $T(v)$–distribution with $v$ degrees of freedom. T–distributions are used in statistics to make inferences on the population variance when the population is assumed to be normally distributed, especially when the population is small.

*Snedecor-Fisher's F–distribution*, $F(v_1, v_2)$, is continuous. If $Y$ and $Z$ are independent $\chi^2$–random variables with $v_1$ and $v_2$ degrees of freedom, respectively, then $X = (Y/v_1)/(Z/v_2)$ has an F–distribution with $(v_1, v_2)$ degrees of freedom. Note that if $X$ is $T(v)$, then $X^2$ is $F(1, v)$. F–distributions are used in statistics when comparing variances of two populations. All these distributions are continuous. Formulas for calculating distribution models and parameters for normal and $\chi^2$–distribution as well as other selected distributions are shown in Table 6.2.

## 6.3 Mathematical Models in Reliability Engineering

To quantitatively describe reliability, the models discussed in Section 6.2 are used by defining specific probabilistic measures. Among the most common are (Barlow and Proschen, 1965, 1975):

- The *probability density function* (PDF) is denoted by *f(t)*. It is a continuous representation of a histogram that shows how the number of component failures is distributed in time. The pdf is always normalized so that its area is equal to 1, as well as scaled to show the probability of failure per unit time.

- The *cumulative distribution function* (CDF), also called the unreliability function or the probability of failure, is denoted by *Q(t)*. It represents the probability that a new component will fail at or before a specified time. The CDF can be calculated by finding the area under the pdf-curve to the left of a specified time, what is described by the formula

$$Q(t) = \int_0^t f(s)ds \qquad (6.9)$$

- On the contrary, if the *unreliability function* is known, the PDF can be obtained as

$$f(t) = \frac{dQ(t)}{dt} \qquad (6.10)$$

- The *reliability function*, also called the survivor function or the probability of success, is denoted by *R(t)*. It represents the probability that a new component will operate longer than a specified time. It can be calculated by finding the area under the pdf to the right of a specified time, what is described by the formula

$$R(t) = \int_t^\infty f(s)ds \qquad (6.11)$$

- In opposition, if the reliability function is known, the PDF can be obtained as

$$f(t) = -\frac{dR(t)}{dt} \qquad (6.12)$$

- The reliability function and the unreliability function satisfy the following equation

$$Q(t) + R(t) = 1 \qquad (6.13)$$

- The *failure rate function*, also called the instantaneous failure rate or the hazard rate, is denoted by $\lambda(t)$. It represents the probability of failure per unit time, $t$, given that the component has already survived to time $t$. Mathematically, the failure rate function is a conditional form of the pdf, and is described by the formula:

$$\lambda(t) = \frac{f(t)}{R(t)} \qquad (6.14)$$

Using the models in equations (6.9) through (6.14) and the probability distributions discussed in Section 6.2, the values of the most important reliability attributes can be determined, which include:

- *Availability (AV)* – ability to be in state to perform the required functions under given work conditions, is described by:
- *Reliability (REL)* – ability to perform the required functions, without failure, for a given time interval, under given work conditions;
- *Maintainability (MAI)* – ability to be retained in, or restored to a state to perform as required, under given conditions of use and maintenance;
- *Maintenance Support Performance (MSP)* – effectiveness of an organization in respect to maintenance support.

Table 6.3 gives selected examples of metrics for these attributes that have been widely used in practice.

**Table 6.3:** Selected examples of metrics for reliability attributes.

| Attributes | Examples of Metrics |
|---|---|
| Availability | mean availability – $A_m$ $(t_1, t_2)$; mean down time – MDT; |
| Reliability | mean failure rate – $\lambda_m$ $(t_1, t_2)$; mean time to failure – MTTF; mean time between failures – MTBF; reliability function – R $(t_1, t_2)$ |
| Maintainability | mean repair time MRT; mean time to restoration – MTTR |
| Maintenance Support Performance | mean administrative delay – MAD; mean logistic delay – MLD |

## 6.4 System Reliability Modelling – The Structural Reliability

To be able to assess the reliability of systems composed of elements with known reliability measures, it is necessary to know the reliability structure of these systems. It can be accomplished through logical and mathematical models that show the functional relationships between all the components that make up the entire system. The reliability of the system is a function of the reliabilities of its components and is presented in graphical form using the so-called *Reliability Block Diagrams* (RBD) (Kapur and Pecht, 2014). RBD-analysis provides a quantitative assessment of basic indicators of system reliability and must be conducted to develop a reliability model of given system. The analysis consists of the following four main steps:

1) Deriving a model in the form of a functional block diagram of the system based on physical principles managing the operations of the system.

2) Defining the logical and topological relationships between functional elements of the system.

3) Determining the conditions under which a particular system may operate in a degraded state, based on performance evaluation studies.

4) Selecting the appropriate spare and repair strategies for maintenance systems.

The model obtained because of the above analysis is the starting point for the calculation of numerical values of specific reliability indices. The reliability block diagram (RBD) is a symbolic way of showing the success or failure combinations for a system under consideration. Some of the practical guidelines for drawing these diagrams are as follows:

• A group of components that are necessary for the performance of the system and/or its mission are drawn in series and connected as a chain (Figure 6.1a).

• Components that can substitute for other components are drawn in parallel and form a redundant structure (Figure 6.1b).

• Each element of the structure has a role analogous to a switch, which if it is in an undamaged state (fitness for use) is closed, and is opened when the component has failed. Any closed path through the diagram is a 'success path'. The failure behaviour of all the redundant components must be specified.

Typical redundancy structures used in practice are:

• Active Redundancy or Hot Standby. A redundant component has the same reliability as the one it replaces.

• Passive Redundancy, Spare, or Cold Standby. The spare element is treated as new until it replaces the failed active element.

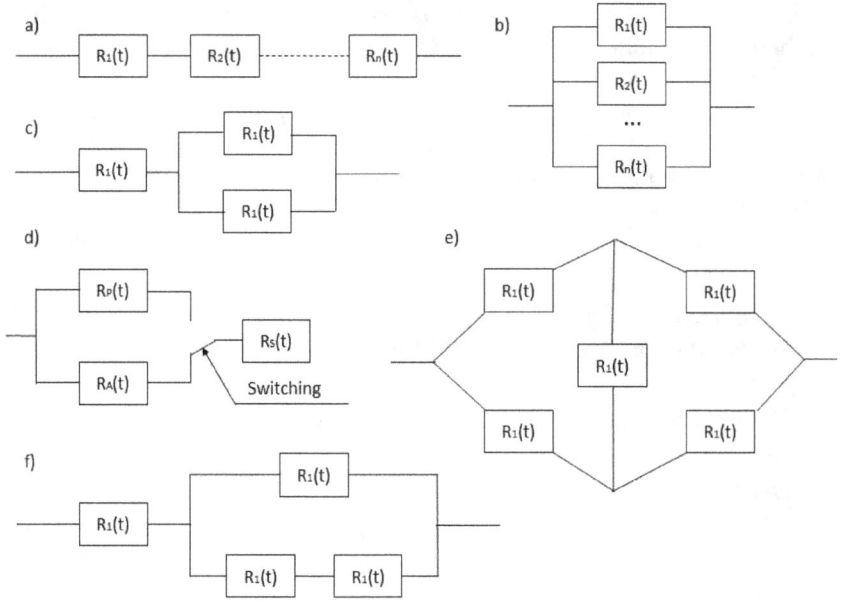

**Figure 6.1:** Reliability blocks diagrams for typical reliability structures: a) serial, b) parallel, c) serial-parallel, d) standby, e) bridge, f) mixed.

- Warm Standby. The standby component has a higher reliability than the operating component but lower than a new one. This is usually a realistic assumption.

### 6.4.1 Series System

In a series system, all its components must operate effectively if the system is to function and operate successfully. This implies that the failure of any component will cause the entire system to fail. The reliability block diagram of a series system is shown in Figure 6.1a. The reliability of each block is represented by $R_i(t)$, and the units need not be physically connected in series for the system to be called a series system. System reliability can be determined using the principles of probability theory. In practice, all elements of the system are assumed to be probabilistically independent. This means that whether one element works does not depend on the state (e.g., failure) of other elements. This assumption, although not always correct, allows to describe the reliability of the system whit the following formula

$$R_s(t) = \Pi_{i=1}^n R_i(t) \tag{6.15}$$

### 6.4.2 Systems with Redundancy

Redundancy consists in the fact that additional elements are implemented in the system, which take over the role of basic elements in a situation when they fail. In the practice of reliability engineering, three types of redundancy are usually used: active, standby, and so-called $(k, n)$ systems. In *active redundancy*, all the parts are operational during the functioning of a system, and all the parts will consume life at the same rate as the individual components. In *standby redundancy*, some parts do not contribute to the operation of the whole system, and they get switched on only if there are failures in the active parts, allowing these parts to run longer than those in active redundancy. There are three main types of standby redundancy: cold, warm, and hot. In *cold standby*, replacement parts are waiting as non-working until they are needed. This significantly extends the expected life of these parts but increases the downtime of the system in the event of an active part failure, and the transient stresses on the parts during switching. In *warm standby*, the replacement parts are usually active but are unloaded. In *hot standby*, the replacement parts form an active parallel system. The lifetime of the hot standby parts are assumed to be consumed at the same degree as active parts.

#### a) Active Redundancy

An active redundant system is a standard 'parallel' system. The system fails only when all components have failed, and it is called a 1-out-of-n or $(1, n)$ system, which implies that at least one out of n subsystems must operate for the system to be functional. Therefore, a series system is a special case of the $(k, n)$ systems. An example of such a structure is shown in Figure 6.1b. The individual elements of the system do not necessarily have to be connected to create such a structure. The system will fail if all the components fail by a given time $t$, or the system will survive the $t$, if at least one of the units survives until $t$. The system reliability can be expressed as follows

$$R_s(t) = 1 - \Pi_{i=1}^{n} [1 - R_i(t)] \tag{6.16}$$

#### b) Standby Systems

A standby system consists of an active unit and one or more inactive (standby) units that become active in the event of the failure of the active unit. The failures of working units ($A$ – active unit) should be signalled by a monitoring subsystem, and then the failed active element is replaced by the spare unit ($P$ – passive unit). The simplest standby configuration is shown in Figure 6.1d. In a general case, there will be $n$ number of units with $(n - 1)$ of them in standby. The reliability for the model shown in Figure 6.1d can be determined ass–ming that the sensing of monitoring system and the switching mechanisms are perfect. In consequence, the second unit is switched on when the first component fails. This means that the system uptime will be the sum of the durability of both elements (*T1* and *T2*), active and passive. When both *T1* and *T2* have the exponential distribution, the active and the standby units have equal constant failure rates, $\lambda$, the reliability function for such a system is given by the equation

$$R_s(t) = e^{-\lambda t}(1 + \lambda t) \tag{6.17}$$

c) $(k, n)$ Systems

A system consisting of n components is called a k-out-of-n and described briefly as $(k, n)$ system if the system only operates when at least k or more components are in an operating state. The reliability model (RBM) diagram for the k-out-of-n system is drawn like the parallel system (Figure 6.1b). When $R_i = R(t)$ for all $I$, and all random variables are independent, the following formula will be obtained:

$$R_s(t) = \sum_{i=k}^{n} \binom{n}{i}[R(t)]^i[1 - R(t)]^{n-i} \tag{6.18}$$

where:

$$1 \leq k \leq n$$

In practice, however, redundancy does not always provide benefits consistent with the above models. This is primarily because common mode failures are caused by phenomena that create dependencies between two or more redundant parts and result in them damaging themselves almost simultaneously. This situation can be caused by many reasons, such as common electric connections, shared environmental stresses, and maintenance problems. So called 'load sharing failures' occur when the failure of one part increases the stress level of other parts. This increased stress level can affect the reliability of the whole system. Therefore, the calculation results obtained from the above formulas should be treated as approximate, and in most cases giving overestimates of system reliability.

# 6.5 References

Azarkhail, M. and Modarres, M. (2007). Markov chain monte carlo simulation for estimating accelerated life model parameters. *The 53rd Annual Reliability & Maintainability Symposium (RAMS).* Orlando, Florida USA, 22–25 January 2007, pp. 214–219.

Azarkhail, M. and Modarres, M. (2012). The evolution and history of reliability engineering: Rise of mechanistic reliability modeling. *International Journal of Performability Engineering,* 8(1): 35–47.

Barlow, R.E. and Proschan, F. (1965). *Mathematical Theory of Reliability.* New York, London, Sydney: John Wiley & Sons, Inc.

Barlow, R.E. and Proschan, F. (1975). *Statistical Theory of Reliability and Life Testing. Probability Models.* NY: Holt, Rinehart, and Winston Inc.

Brooks, S.P. (1998). Markov Chain Monte Carlo method and its application. *The Statistician, Royal Statistical Society,* 47(1): 69–100.

Bukowski, L. (2019) *Reliable, Secure and Resilient Logistics Networks. Delivering Products in a Risky Environment.* Springer Nature Switzerland AG 2019. ISBN 978-3-030-00849-9 (Hardcover), ISBN 978-3-030-00850-5 (eBook).

Congdon, P. (2003) *Applied Bayesian Modeling.* New York: Wiley Series in Probability and Statistics. John Wiley & Sons.

Denson, W. (1998). The History of Reliability Prediction. *IEEE Transactions on Reliability,* 47(3): 321–328.

Ebel, G.H. (1998). Reliability physics in electronics: A historical view. *IEEE Transactions on Reliability,* 47(3): 379–389.

Fleming, K.N. and Mosleh, A. (1985). Common-cause data analysis and implications in system modeling. *Proc. Int'l Topical Meeting on Probabilistic Safety Methods and Applications,* 1(3): 1–12.

Fragola, J.R. (1996). Reliability and risk analysis data base development: An historical perspective. *Reliability Engineering and System Safety,* 51(2): 125–136.

Hald, A. (1999). On the maximum likelihood in relation to inverse probability and least squares. *Statistical Science,* 14(2): 214–222.

Joyce, D. (2016). *Common Probability Distributions Probability and Statistics.* https:// mathcs. clarku.edu/~djoyce/ma218/distributions.pdf.

Kapur, K.C and Pecht, M. (2014). *Reliability Engineering.* John Wiley & Sons.

Keller, W. and Modarres, M. (2005). A historical overview of probabilistic risk assessment development and its use in the nuclear power industry: A tribute to the Late Professor Norman Carl Rasmussen. *Reliability Engineering and System Safety,* 89(3): 271–285.

Montgomery, D.C. and Runger, G.C. (2003). *Applied Statistics and Probability for Engineers.* New York: John Wiley & Sons, Inc.

Nelson, W.B. (1990). *Accelerated Testing: Statistical Models, Test Plans, and Data Analysis.* New York: Wiley Series in Probability and Statistics, John Wiley & Sons.

O'Connor, P.D.T. and Kleyner, A. (2012). *Practical Reliability Engineering,* Fifth Edition, John Wiley & Sons, Ltd.

Pecht, M. and Dasgupta, A. (1995). Physics-of-failure: An approach to reliable product development. *Journal of the Institute of Environmental Science,* 38(5): 30–34.

Ross, S.M. (2004). *Introduction to Probability and Statistics for Engineers and Scientists.* Mass., USA: Elsevier Academic Press.

Sage, A.P. and Rouse, W.B. (1999). *Handbook of Systems Engineering and Management.* New York: John Wiley & Sons Inc.

Scheaffer, R.L., Mulekar, M. and McClave, J.T. (2011). *Probability and Statistics for Engineers.* Brooks/Cole, Cengage Learning.

Smith, D.J. (2011). *Reliability, Maintainability and Risk Practical Methods for Engineers.* Maas., USA: Elsevier Ltd.

Soong, T.T. (2004). *Fundamentals of Probability and Statistics for Engineers.* New York: John Wiley & Sons Ltd.

Wasserman, L. (2005). *All of Statistics. A Concise Course in Statistical Inference.* New York: Springer Science+Business Media.

# 7

# Safety Engineering
## The Concept of Effective Protection*

Chapter 7 analyses the development of safety issues from the dawn of time to the present day. It presents the most important definitions and assumptions on which the foundations of safety science were built. Three most important trends in the construction of models of hazard and accident occurrence were discussed, namely, simple linear accident models, complex linear accident models, and complex non-linear accident models. The advantages and disadvantages of models based on metaphors such as the domino effect, Swiss cheese, and epidemiological model are presented. The second part of the chapter presents the most important Safety Engineering methods and tools. The limitations of using classical models based on the principle of causality in practice were analysed, and modern complex models were presented against this background. It focuses on two different approaches to the safety of complex systems, namely the Systems-Theoretic Accident Mode and Processes/Systems Theoretic Process Analysis (STAMP/STPA) and Functional Resonance Accident Model (FRAM). Their characteristics and basic advantages are presented. Areas of practical applications and development trends are indicated.

## 7.1 The Origins and Development of the Safety Science – Accident Models

An accident is usually defined as: "a short, sudden and unexpected event or occurrence that results in an unwanted and undesirable outcome …

---

* https://orcid.org/0000-0002-2630-3507.

and must directly or indirectly be the result of human activity rather than a natural event" (Hollnagel, 2004). Accident prevention is the most basic of all safety management paradigms; therefore, understanding how accidents occur is fundamental to establishing interventions to prevent their occurrence. For this reason, researchers and practitioners have been searching for nearly 100 years for conceptual models that would capture the mechanisms of hazards and accidents. Various models were created, such as linear models which suggest one factor leads to the next and to the next leading up to the accident, and complex nonlinear models which assume multiple factors are acting concurrently and by their combined influence, lead to accident occurrence. This diversity was because:

Accident models affect the way people think about safety, how they identify and analyse risk factors and how they measure performance … they can be used in both reactive and proactive safety management … and many models are based on an idea of causality … accidents are thus the result of technical failures, human errors or organizational problems (Hovden et al., 2010).

The history of accident models to date can be traced from the 1920s through three distinct phases: Simple linear models, Complex linear models, and Complex nonlinear models. (Hollnagel, 2010.) Each type of model is supported by specific assumptions:

- *Simple linear models* assume that accidents are the culmination of a series of events or circumstances which interact sequentially with each other in a linear way and thus accidents are preventable by eliminating one of the causes in the linear sequence.

- *Complex linear models* assume that accidents are a result of a combination of unsafe acts and latent hazard conditions within the system which follow a linear pathway. The factors furthest away from the accident are attributed to actions of the organization or environment and factors at the sharp end being where humans ultimately interact closest to the accident; it means that accidents could be prevented by focusing on establishment barriers and defenses.

- *Complex nonlinear models*. A new way of thinking about accident modelling has moved towards recognizing that accident models need to be nonlinear; that accidents may be modelled as resulting from combinations of mutually interacting variables which occur in real world environments, and it is only through understanding the combination and interaction of these multiple factors that accidents can truly be understood and prevented (Hollnagel, 2010).

### 7.1.1 Simple Sequential Linear Accident Models

Heinrich's Domino Theory

The first sequential accident model was the 'Domino effect' (Heinrich, 1931). The model was based in the assumption that: "the occurrence of a preventable injury is the natural culmination of a series of events or circumstances, which invariably occur in a fixed or logical order ... an accident is merely a link in the chain". This model suggested that certain accident factors could be described as lined up sequentially like dominos. Heinrich proposed that: "an accident is one of five factors in a sequence that results in an injury ... an injury is invariably caused by an accident and the accident in turn is always the result of the factor that immediately precedes it. In accident prevention the bull's eye of the target is in the middle of the sequence – an unsafe act of a person or a mechanical or physical hazard". Heinrich's five factors were:

- Social environment (ancestry),
- Fault of the person,
- Unsafe acts, mechanical and physical hazards,
- Accident, and
- Injury.

Based on the domino metaphor, an accident was considered to occur when one of the dominos or accident factors falls and has an ongoing knockdown effect ultimately resulting in an accident.

Loss Causation Model

Bird and Germain (1985) developed an updated domino model which reflected the direct management relationship with the causes and effects of accident loss and incorporated arrows to show the multilinear interactions of the cause-and-effect sequence. This model became known as the Loss Causation Model and was represented by a line of five dominos, linked to each other in the form of a chain. In this chain, individual domino blocks represented the following factors:

- Lack of control – due to inadequate program, standards, or compliance to requirements,
- Basic causes – due to personal, or job factors,
- Immediate causes – because of standards acts and conditions,
- Incident – because of contact with energy or substance, and
- Loss – concerning people, property, and processes.

### 7.1.2  Complex Linear Accident Models

Sequential models focus on the view that accidents happen in a linear way where event A leads to event B which leads to event C and so on. They examine the chain of events between multiple causal factors presented in a sequence usually from left to right. Accident prevention methods developed based on these sequential models focus on search for root causes and eliminating them or putting in place barriers to deny access for threats. In the 1970s, sequential accident models had begun to incorporate multiple events in the sequential chain. Key models developed in this period include primarily energy damage models, time sequence models, epidemiological models, and systemic models.

Energy Damage Models

The Energy Damage Models assume that "Damage is a result of an incident energy whose intensity at the point of contact with the recipient exceeds the damage threshold of the recipient" (Viner, 1991). According to these models, hazards are the sources of potentially damaging energy, and an accident, injury or damage may result from the loss of control of the energy when there is a failure of the hazard control mechanism. Such mechanisms may include physical or structural containment, barriers, processes, and procedures. The space transfer mechanism is how the energy and the recipient are brought together assuming that they are initially distant from each other. The recipient boundary is the surface that is exposed and vulnerable to the energy (Viner, 1991).

Time Sequence Models

Benner (1975) identified four main issues which were not addressed in the basic type of domino model:

- defining a beginning and end of an accident;
- representing the events that happened on a sequential time line;
- the need for a structured method for discovering the relevant factors involved; and
- using a charting method to define events and conditions.

Viner's Generalized Time Sequence Model is an example of a time sequence model that addresses these four requirements. He considers that the structure for analysing the events in the occurrence-consequence sequence provided by this kind of models draws attention to counter measures that may not be evident. In the first time zone, there is the opportunity to prevent the event occurring, because there is some time between the event initiation and the event. Second time zone offers a warning of the impending existence of an event mechanism and the opportunity to take remedial action to reduce the likelihood of the event. The last time zone gives an opportunity to influence the outcome and protect the exposed groups (Viner, 1991).

## Epidemiological Models

Epidemiological Accident Models are derived from the study of disease epidemics and the search for causal factors around their development. Gordon (1949) recognized that "injuries, as distinguished from disease, are equally susceptible to this approach", which means that our understanding of accidents would benefit by assuming that accidents are caused by: "a combination of forces from at least three sources, which are the host – and man is the host of principal interest – the agent itself, and the environment in which host and agent find themselves." (Gordon, 1949). Benner (1975, 1984) moved away from identifying a few causal factors to understanding how multiple factors within a system combined. These models assume that an accident combines agents and environmental factors which influence a host environment (like an epidemic) that have negative effects on the organism or an organization.

Reason used the epidemiological metaphor to present the idea of 'resident pathogens' when emphasizing: "the significance of causal factors present in the system before an accident sequence actually begins ... and all man-made systems contain potentially destructive agencies, like the pathogens within the human body" (Reason, 1987). The term became known as 'latent errors', then changed to 'latent failures' evolving further when the notion 'latent conditions' became preferred.

## Systemic Models

By the 1980s safety researchers have stated that previous accident models did not reflect any realism as to the true nature of the observed accident phenomenon. Benner noted that "one element of realism was nonlinearity ... models had to accommodate nonlinear events. Based on these observations, a realistic accident model must reflect both a sequential and concurrent nonlinear course of events, and reflect events interactions over time" (Benner, 1984). This opinion was supported by Rasmussen, who acknowledged that the identification of events and causal factors in an accident are not isolated but "depend on the context of human needs and experience in which they occur and by definition ... therefore will be circular" (Rasmussen, 1990). Systemic accident models examined the idea that systems failures, rather than human failure, were a major contributor to accidents (Hollnagel, 2004), and recognized that events do not happen in isolation of the systemic environment in which they occur.

Further works by Rasmussen (1982, 1986) and Reason (1979, 1984a, 1984b, 2008) have confirmed that the skill-based, rule-based, and knowledge-based distinctions of human error in operations are very significant. Rasmussen wrote extensively on the problem of causality in the analysis of accidents introducing concepts on the linkage between direct cause-effect, timeline and accident modelling. He recognized that socio-technical systems were both complex and unstable, as well as a flow of events does not consider "closed loops of interaction among events and conditions at a higher level of individual and organizational adaption … with the causal tree found by an accident analysis is only a record of one past case, not a model of the involved relational structure" (Rasmussen, 1990). The model developed by Reason on this basis is known as the Swiss Cheese Model (see Figure 7.1).

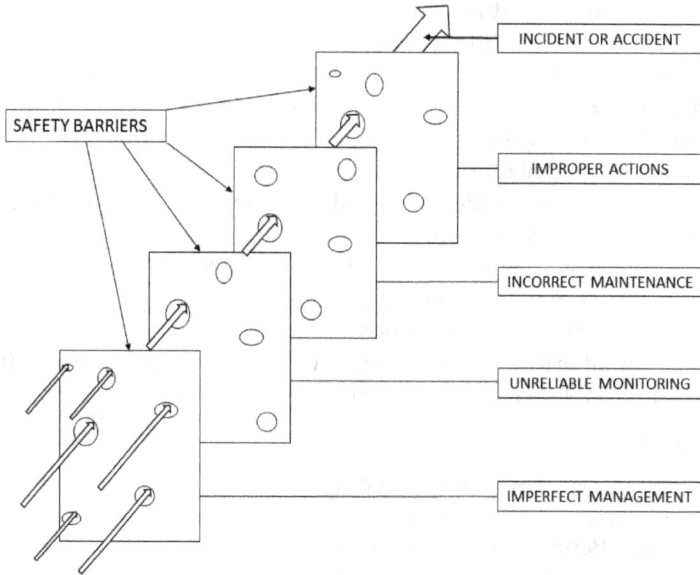

**Figure 7.1:** Example of graphical interpretation of the Swiss Cheese Model.

### 7.1.3 Complex Nonlinear Accident Models

In the 1980s, Charles Perrow's book entitled *Normal Accidents: Living with High-Risk Technologies* was published, which started a new trend in the way safety problems were viewed. He argued that technological advances had made systems not only closely coupled but very complex, so much so that he termed accidents in these systems as being a natural state in the real world. Perrow's theory postulated that tightly coupled systems had low level of tolerance for even the smallest disturbance which would result in adverse outcomes. Thus, strongly coupled systems were inherently unsafe, so that operator errors are inevitable due the way the system parts were interrelated (Perrow, 1984, 2011). The individual elements of complex systems are connected in different ways, so that in many cases it is almost impossible to understand their relationships and interactions. (Perrow, 1984). The idea of 'normal catastrophes' has resonated greatly and has resulted in new trends in systems safety modelling. Two new accident models were introduced in the early 2000s which referred to earlier models of the type of linear accident models (Hovden et al., 2010): the Systems-Theoretic Accident Model and Process (STAMP) (Leveson, 2004), and the Functional Resonance Accident Model (FRAM) (Hollnagel, 2004).

Systems-Theoretic Accident Model and Process (STAMP)

Leveson's model considered systems as "interrelated components that are kept in a state of dynamic equilibrium by feedback loops of information and control" (Leveson, 2004). It emphasized that safety management systems were required to continuously control tasks and impose constraints to ensure system safety. The model of accident investigation focused on the problem because the controls that were in place were not able to detect and prevent changes that finally lead to an accident. Leveson developed a classification of flaws method to assist in identifying the factors which contributed to the event, and which pointed to their place within a looped and linked system. This model expands on the barriers and defences approach to accident prevention and is tailored to proactive preventive actions. The STAMP model has become one of the foundations of Safety Engineering and will be discussed in detail in Section 7.2.

Functional Resonance Accident Model (FRAM)

Erik Hollnagel introduced the concept of a three-dimensional way of thinking about accidents in highly complex and tightly coupled socio-technical systems. He established the goals of organizations as moving from putting in place barriers and defences to focusing on systems able to monitor and control any variances, and to be human errors tolerant. FRAM is based on complex systemic accident theory but considers that system variances result in an accident only when the system is unable to tolerate such variances in its normal functioning mode. System variance is recognized as normal within most systems and represents the necessary performance variability needed for complex systems to operate, including limitations of design, imperfections of technology, work conditions, and combinations of inputs which generally allowed the system to work (OHS, 2013). Four main sources of variability were identified as: humans, technology, latent conditions, and barriers (Hollnagel, 2010).

Figure 7.2 shows an example of the FRAM model interpretation for a certain system, described by many parameters subjected to random

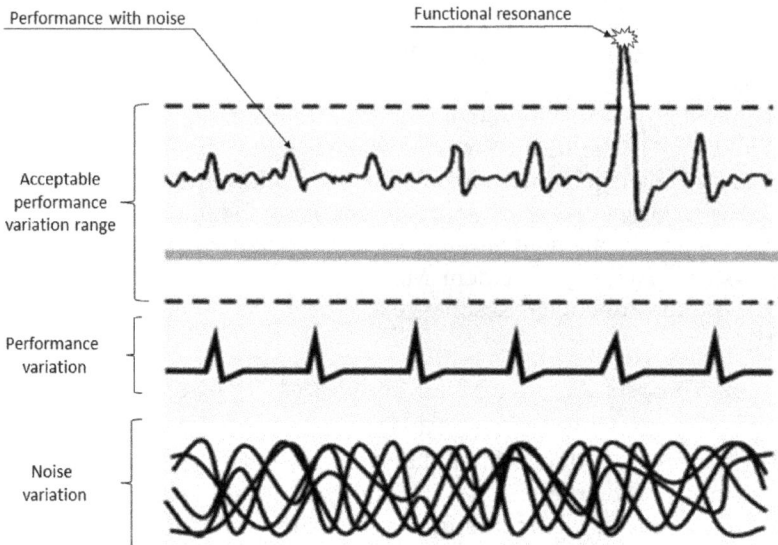

**Figure 7.2:** An example of the FRAM model interpretation (based on Hollnagel, 2010).

changes. The behaviour of the entire system (presented in the form of a time course in the upper part of the figure) is a resultant of the instantaneous values of individual system parameters and the impact of the environment (in the drawing presented by noise variation). A characteristic feature of functional resonance phenomena is that a particularly unfavorable (and very rarely occurring) configuration of system parameters and ambient values can lead to a significant exceeding of the allowable parameter describing the entire system, while its individual and surrounding parameters remain constant within the limits of acceptable changes. The FRAM model has become one of the bases of Safety Engineering and will be discussed in detail in Section 7.2.

To summarize this section, it can be said that there are six main uses for accident causation models (Hovden et al., 2010):

- create a common understanding of accident phenomena through a shared simplified representation of real-life accidents,
- help understand and communicate risk problems,
- give a basis for inter-subjectivity, therefore preventing personal biases regarding accident causation and providing an opening for a wider range of preventive measures,
- guide investigations regarding data collection and accident analyses,
- support analyse interrelations between factors and conditions,
- use different accident models highlighting different aspects of processes, conditions, and causes.

## 7.2  Safety Engineering Methods and Tools

Parallel to the research in the field of safety definition and modelling, which became the basis for the birth of a new discipline – Safety Science – in engineering practice, more and more attention began to be paid to the safety of technical systems around the 1940s. In the early days of Safety Engineering, the 'trial-and-error' methodology became dominant in the aviation field and was called the 'fly-fix-fly' approach. As aviation grew as an increasingly popular passenger transport service, this approach became too hazardous and the risk of an accident unacceptable. In the area of defence, with the introduction of weapons of mass destruction, as well as in the rapidly expanding space travel, security requirements have become a priority, leading to a change in approach from the traditional 'trial-and-error" to the new one, namely 'first-time safe' effort. The first result of this pressure for change was the development in the 1960s of safety requirements for systems and their components by the U.S. Air Force Ballistic System Division, which became the basis for MIL-STD-882, published in 1969. This document and its subsequent modifications MIL-STD-882A and MIL-STD-882B have become the fundamental knowledge base for systems safety, particularly for DOD. In later years, other institutions such as NASA and the aerospace industry (both military and civilian) developed their own documents modelled after MIL-STD-882 (Stephans, 2004).

The next phase in the development of Safety Engineering was related to the energy industry, especially nuclear power. The result of increased risk awareness in these areas was the publication by the Atomic Energy Commission (AEC) in 1973 of a manual entitled *Management Oversight and Risk Tree* (MORT). International cooperation on nuclear power safety has resulted in the INES (International Nuclear and Radiological Event Scale) as the safety pyramid model shown in Figure 7.3. In the 1980s, as systems became more complex and expensive, interest in the practical aspects of safety assurance extended to other areas of technology, such as the chemical industry. In 1985, the American Institute of Chemical Engineers developed the Guidelines for Hazard Evaluation Procedures, identified later as Hazard and Operability Programs (HazOp). This trend has been called Facility System Safety. In the 1990s, there was a shift from a systems approach (emphasis on device structure) to a process approach (emphasis on runs and system behaviour). The result was the emergence of the O&SHA (Operating and Support Hazard Analysis) process safety regulation, which required that "the risk associated with a manufacturing or processing site with listed substances be assessed and appropriate actions be taken to mitigate the results of an accident to protect the workers" (Stephans, 2004). The beginning of the 21st century has seen a

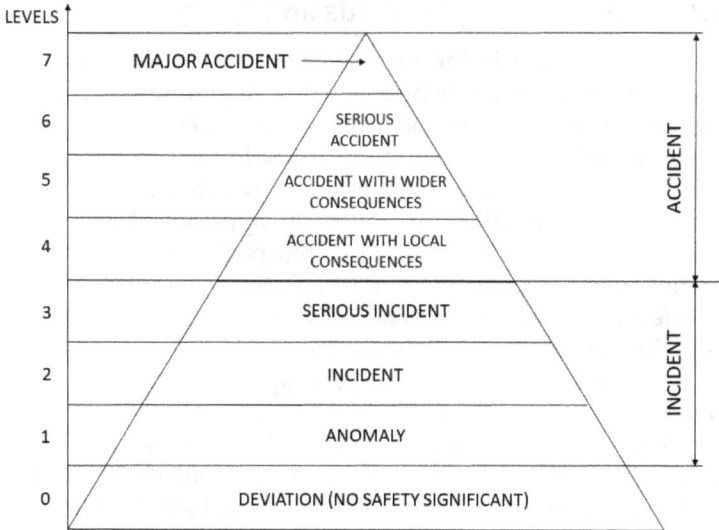

**Figure 7.3:** The safety pyramid model INES – International Nuclear and Radiological Event Scale [based on (IAEA, 2014)].

shift towards a holistic approach, aiming to integrate Safety Engineering with other areas of expertise, such as Automation and Computer Science, and aiming to achieve system safety by building into the system already at the design stage (so-called *intrinsic safety*).

Most safety models used in engineering practice are built on the assumption that accidents are the result of a chain of consecutive events. Such models are useful in cases caused by failures of physical components and for relatively simple systems. However, in relation to currently used systems, these assumptions are not fulfilled, mainly for the following reasons (Leveson, 2011):

- Fast pace of technological change. New technologies, especially high-tech solutions, are changing faster than the engineering techniques to cope with new challenges and the risks they pose. Experience learned over decades about designing to prevent accidents become ineffective when older technologies are replaced with modern ones. New technology introduces us into hitherto unexplored and sometimes even so-called unknown-unknown areas. Additionally, the time to market for new products has significantly decreased and strong pressures exist to decrease this time even further. This means that there is generally a lack of time for carefully testing new systems and designs to understand all the potential hazards and risks before commercial use.

- Changing nature of accidents. Information and telecommunication technologies have created a revolution in most fields of engineering, but system engineering and system safety engineering techniques have not kept up with these changes. Cyber-physical systems have also changed the nature of accidents. Many of the approaches to prevent accidents that worked successfully on electromechanical components (such as redundancy of components to protect against individual component failure) became ineffective in controlling accidents that arise from the use of digital systems, both in terms of hard and software.

- Emergence of new types of hazards. The most common accident models assume that accidents are the result of an uncontrolled and undesired release of energy or interference in the normal flow of energy. Increasing levels of process automation are driving the need for more and more information, making system security even more dependent on imperfect data often acquired in real time. It means that in many cyber-physical systems the software is safety-critical and plays an increasingly important role in accidents.

- Increasing complexity and coupling. Currently, systems are being designed and implemented that are characterized by an interactive complexity that cannot be thoroughly planned, understood, anticipated, or protected against. A serious problem is that such sophisticated systems are created that the control and management of them exceeds the capabilities of their users, especially in abnormal situations.

- Decreasing tolerance for single accidents. The increased complexity of systems and processes has resulted in an increase in the potential financial loss associated with system failure, as well as an increase in the number of people at risk of loss of life or health. Therefore, learning from accidents using the 'fly-fix-fly' approach to safety needs to be supplemented and supported with increasing importance on preventing the first one.

- More complex relationships between humans and automation. Humans are increasingly sharing management and decision-making with automation implementing the decisions. These changes are leading to new types of human error, and changes in the frequency of occurrence of each type of error, e.g., increasing errors of omission versus commission. Human behaviour is influenced by the context in which it occurs, and operators in high-tech systems are often limited in their actions by the automation systems built into the systems they control. Also, ambiguities occurring in communication between humans and machines is becoming an increasingly important factor in accidents.

- Changing regulatory and public views of safety. In today's complex and interrelated societal structure, responsibility for safety should be shifted from the individual humans and companies (safety management) to government (safety governance). Individuals no longer have the possibility to control the systemic risks and are demanding that government undertake greater responsibility for controlling hazards through laws and various forms of oversight and regulation. These changes are challenging both our accident models and the accident prevention techniques based on them.

To overcome these limitations of classical safety analysis methods, new concepts have been developed, of which two approaches deserve special attention, namely: STAMP/STPA methodology and FRAM model. Both these will be briefly discussed below.

Systems-Theoretic Accident Mode and Processes/Systems Theoretic Process Analysis (STAMP/STPA) is an analytical method proposed by Nancy G. Leveson of the Massachusetts Institute of Technology for problems caused by interactions between physical system and social infrastructure areas. Conventional safety analysis methods, such as fault tree analysis (FTA) or failure mode and effect analysis (FMEA), available since the 1960s, are convenient for analysing single failures in technical equipment or organizations but have limits for application to complicated complex systems under continuous changes and variability. The reason is that accidents occur in complex systems due to failures of single components and inter-component communication errors. The STAMP/STPA method is based on a top-down process and on the concept of big picture analysis of interactions caused by intra-system components to control accidental emergent properties and prevent accidents (Yoshida, 2021).

The process of implementing the method follows the following four steps:

Step 1: Separate incidents, accidents, hazards, and safety constraints. Define the accidents and hazards to be analysed in a system. Accidents are defined in a broad sense as any unacceptable loss of some significant values. Hazard is understood as a threatening condition that should be removed or neutralized as quickly as possible. Then, finally, recognize and identify system safety constraints for hazard control.

Step 2: Design the control structure. Build a control structure by analysing the system's parts (subsystems, equipment, organizations, etc.) together with component interactions, and establish the safety constraints.

Step 3: Identify unsafe control actions (UCA). Determine controller-issued inter-component instructions (e.g., control actions) necessary to activate safety constraints. Identify unsafe control actions. Four rules are recommended to help identify unsafe control actions:

- Not providing causes hazard. Failure to provide any of the controls required for safety leads to a hazard.
- Providing causes hazard. An unsafe control action is provided which leads to a hazard.
- Too early, too late, or a wrong order causes a hazard. A usually safe control action provided too late, too early, or out of sequence leads to a hazard.
- Stopping too soon or applying too long causes hazard. A safe control action stopped too soon or applied too long leads to a hazard.

Step 4: Identify hazard causal factors. For each UCA identified in Step 3, distinguish the relevant controllers, and control target processes and create a control loop diagram (CLD) by referring to an adopted model of cause-and-effect scenario generation to identify hazard causal factors. Intended use of CLD is to focus consideration on few most important components rather than to consider multiple component interactions simultaneously. It contains a list of common causes of the occurrence of hazard causal factors.

A comprehensive example of applying the STAMP method to analyse a water contamination accident is shown in the paper (Leveson et al., 2004).

The basis for accident analysis using the Functional Resonance Accident Model (FRAM) is to first define the functional entities that are of importance for the given scenarios or tasks. The entities are more likely to be characteristic or recurrent functions than to be system structures or physical units. These functional entities are described in terms of the following relations (Hollnagel, 2018):

- Inputs (I) – variables needed to perform the function. Inputs constitute the links to previous functions and can be either transformed or used by the function to produce the outputs.
- Outputs (O) – variables produced by the function. Outputs constitute the links to subsequent functions.
- Resources (R) – assets needed by the function to process the inputs (in terms of, e.g., hardware, procedures, software, energy, manpower) into outputs.
- Controls (C) – constraints, that serve to supervise or restrict the function (to monitor it and adjusts if it does not meet the requirements). Controls can be active functions or plans, procedures, standards, and guidelines.

- Preconditions (P) – system conditions that must be fulfilled before a function can be carried out. A typical precondition may be a requirement that some condition must be met before you can proceed to the next step.
- Time (T) – a special kind of resource. All processes take place in time and are managed by time. Time can also act as a constraint, e.g., time can also act as a constraint, e.g., if a certain time window (duration) is set for an activity.

The graphical representation of the function model is a hexagon, each vertex of which corresponds to one unit from the list above. A model of a complex process is a set of individual hexagons connected in an appropriate way, e.g., as shown in Figure 7.4 for a process consisting of four functions.

The FRAM methodology is based on four main principles, namely (Hollnagel, 2018):

- First principle: the equivalence of successes and failures. The main premise of the FRAM method, as opposed to traditional methods, is the idea that "things that go right and things that go wrong happen in much the same way". The fact that the results of these actions are diametrically opposed does not necessarily mean that their mechanisms are also different.

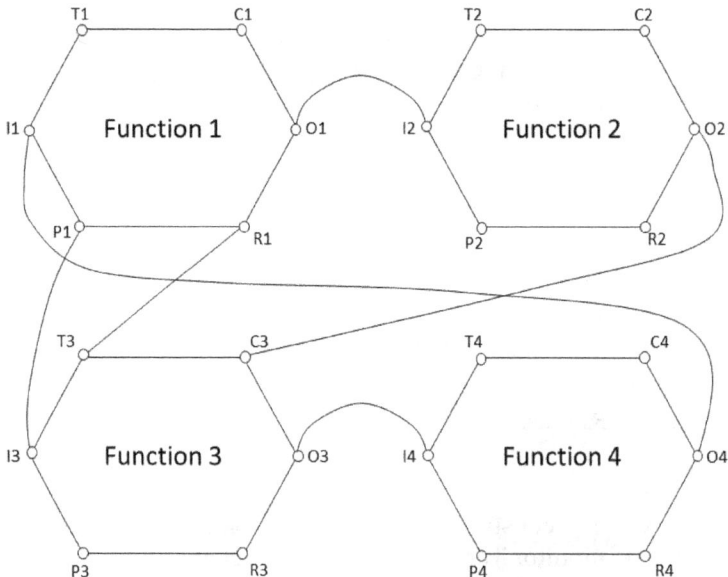

**Figure 7.4:** An example of a complex process model built using the FRAM method.

- Second principle: approximate adjustments. Complex socio-technical systems require different specifications and modelling methods than simple technical systems. Humans are not machines and cannot be modelled like the non-living components of systems. This means that individual model parameters need to be adjusted for specific resource types and current conditions. Adjustments should be made by individuals, groups, and organizations at all levels.

- Third principle: emergent outcomes. Both acceptable and unacceptable results can be explained as emergent outcomes resulting from variability rather than because of single or multiple cause-effect event chains. It is rare that the variability of a single function is so large that it alone leads to an error, a failure, or an accident.

- Fourth principle: functional resonance. Variation in functions can produce different effects, e.g., enhancing or weakening the result in a way that is not proportional to the variation. This effect can be thought of as analogous to the phenomenon of resonance in dynamic physical systems.

The FRAM model has many practical implementations and there are many examples in the literature and on the Internet of using it for security analysis of systems and in other areas related to risk management (e.g., Hollnagel, 1993, 1998, 2004, 2005, 2010, and 2018).

In summary, it can be said that in recent years there has been a shift in emphasis in Safety Engineering from the concept of so-called Safety-I (freedom from unacceptable harm) to Safety-II (focus on how things can go well). This means that the distinction between classical Safety Engineering and Resilience Engineering, which is currently developing very rapidly, is becoming blurred. This problem will be discussed in detail in the following chapters.

## 7.3 References

Benner, L. (1975). Accident investigations: Multilinear events sequencing methods. *Journal of Safety Research*, 7(2): 67–73.

Benner, L. (1984). Accident models: How underlying differences affect workplace safety. Paper presented at the *International Seminar on Occupational Accident Research*, Saltjobaden, Sweden.

Bird, F. and Germain, G. (1985). *Practical Loss Control Leadership*. Loganville, Georgia: International Loss Control Institute, Inc.

Gordon, J. (1949). The epidemiology of accidents. *American Journal of Public Health*, 39: 504–515.

Heinrich, H.W. (1931). *Industrial Accident Prevention: A Scientific Approach*. New York: McGraw-Hill.

Hollnagel, E. (1993). *Human Reliability Analysis: Context and Control*. London: Academic Press.

Hollnagel, E. (1998). *Cognitive Reliability and Error Analysis Method: CREAM*. MA, USA: Elsevier.

Hollnagel, E. (2004). *Barriers and Accident Prevention*. Aldershot: Ashgate.

Hollnagel, E. (2005). *The ETTO Principle: Efficiency-Thoroughness Trade Off*. Retrieved from http://www.ida.liu.se/~eriho/ETTO_M.htm.

Hollnagel, E. (2010). *FRAM Background*. Retrieved from http://sites.google.com/site/erikhollnagel2/coursematerials/FRAM_background.pdf.

Hollnagel, E. (2018). *Safety-II in Practice*. UK: Routledge.

Hovden, J., Abrechtsen, E. and Herrera, I.A. (2010). Is there a need for new theories, models, and approaches to occupational accident prevention? *Safety Science*, 48(8): 950–956.

IAEA. (2014). *The Use of the International Nuclear and Radiological Event Scale (INES) for Event Communication*. https://www-pub.iaea.org/MTCD/Publications/PDF/INES_web.pdf.

Leveson, N. (2004). A new accident model for engineering safer systems. *Safety Science*, 42: 237–270.

Leveson, N. (2011). *Engineering a Safer World. Systems Thinking Applied to Safety*. Cambridge, MA, USA: The MIT Press.

Leveson, N., Daouk, M., Dulac, N. and Marais, K. (2004). *Applying STAMP in Accident Analysis*. Cambridge, MA, USA. MIT. https://shemesh.larc.nasa.gov/iria03/p13-leveson.pdf.

Perrow, C. (1984). *Normal Accidents: Living with High-Risk Technologies*. New York: Basic Books Inc.

Perrow, C. (2011). *The Next Catastrophe: Reducing Our Vulnerabilities to Natural, Industrial, and Terrorist Disasters*. Princeton, NJ: Princeton University Press.

Rasmussen, J. (1982). Human errors: A taxonomy for describing human malfunction in industrial installations. *Journal of Occupational Accidents*, 4: 311–335.

Rasmussen, J. (1986). *Information Processing and Human-Machine Interaction*. Amsterdam: North-Holland.

Rasmussen, J. (1990). Human error and the problem of causality in analysis of accidents. Paper presented at the Human Factors in Hazardous Situations. *Proceedings of a Royal Society Discussion Meeting, 28–29 June 1989, UK*.

Reason, J.T. (1979). Actions not as planned: The price of automatization. *In*: Underwood, G. and Stevens, R. (Eds.). *Aspects of Consciousness* (Vol. 1). London: Wiley.

Reason, J.T. (1984a). Lapses of attention. *In*: Parasuraman, R. and Davies, R. (Eds.). *Varieties of Attention*. New York: Academic Press.

Reason, J.T. (1984b). Absent-mindedness and cognitive control. *In*: Harris, J. and Morris, P. (Eds.). *Everyday Memory, Actions and Absent-Mindedness*. London: Academic Press.

Reason, J.T. (1987). The Chernobyl Errors. *Bulletin of the British Psychological Society*, 40: 201–206.

Reason, J.T. (2008). *The Human Contribution: Unsafe Acts, Accidents and Heroic Recoveries*. Surrey: Ashgate.

Stephans, R. (2004). *System Safety for 21st Century*. Hoboken, New Jersey: John Wiley & Sons, Inc.

Viner, D. (1991). *Accident Analysis and Risk Control*. Melbourne: Derek Viner Pty Ltd.

Yoshida, K. (2021). Introduction of System Safety Analysis Method (STAMP/STPA) in the development of the PCB Inspection System. *OMRON TECHNICS* Vol. 53.006EN 2021.5.

# 8

# Security Engineering
## The Concept of Cyber-security<sup>*</sup>

Chapter 8 addresses the problems of ensuring the security of complex systems, particularly cyber systems. The introductory section discusses the concepts of *trustworthy secure systems,* which create the foundation for functioning as intended also under the conditions of disruptions, hazards, and threats, with respect to given constraints, limitations, and uncertainty. The following section presents the idea of the lifecycle-based System Security Engineering, which refers to all processes and activities associated with the system throughout its entire life, with the focus on specific security considerations. Based on these assumptions, the systems security engineering framework was developed, which provides a conceptual view of the key contexts of the systems security engineering activities, both technical and nontechnical. The next section focuses on cyber security as a process for protecting information by preventing, detecting, and responding to attacks. The cyber security concept called CIA triad is discussed, and in conclusion, a general model for securing resources from adversarial attacks through appropriate security policies and selection of appropriate countermeasures is presented.

## 8.1 The Principles and Concepts of the Security Engineering

### 8.1.1 Basic Concepts of Security Engineering

The concept of *trustworthy secure systems* derives from a wide variety of stakeholder needs that are determined by mission, business, and a

---

\* https://orcid.org/0000-0002-2630-3507.

spectrum of other objectives and concerns. The characteristics of these systems include an evolving growth in the geographic size, number, number of components and technologies that compose the systems. The complexity and dynamicity in the interactions, behaviour, and the increased mutual dependence resulting sometimes in catastrophic loss due to disruptions, hazards, and threats within the global operating environment (NIST, 2018). Managing the complexity of today's systems and be able to provide them with the required level of security means there must be a high confidence in the feasibility, regarding the ability of a system to function as securely as intended. This basis provides the foundation that the system functions as intended also under the conditions of disruptions, hazards, and threats, with respect to given constraints, limitations, and uncertainty. Systems engineering, and system-of-systems engineering provide the foundation for a disciplined approach to creating trustworthy secure systems.

*Trustworthiness*, specifically in this context means: "… worthy of being trusted to fulfil whatever critical requirements may be needed for a particular component, subsystem, system, network, application, mission, enterprise, or other entity" (Neumann, 2004). In general trustworthiness requirements can include many attributes such as safety, security, reliability, dependability, performance, resilience, and survivability under a wide range of potential adversity in the form of disruptions, hazards, and threats. From a security perspective, a trustworthy system is a system that meets specific security requirements in addition to meeting other critical requirements. Systems security engineering provides the needed complementary engineering capability to design and deliver trustworthy secure systems. These systems are less susceptible to the effects of different threats that includes attacks arranged by an intelligent adversary. While it is impossible to predict all potential forms of adversity and to stop all anticipated hazards, and threats, the basic architecture and design of systems can make those systems inherently less vulnerable, provide an increased level of robustness, offering fault-tolerance and resilience that can be leveraged by system owners and operators.

The term *security* is usually defined as "the freedom from those conditions that can cause loss of assets with unacceptable consequences" (Anderson, 1972; Bishop, 2005; Levin et al., 2007). The notion *asset* refers to an item of value to stakeholders. An asset may be tangible (e.g., hardware, firmware, computing platform, network device, or other technology component) or intangible (e.g., data, information, software, trademark, copyright, patent, intellectual property, image, or reputation). The value of an asset is driven by the stakeholders in consideration of life cycle concerns that include, but are not limited to, those concerns of business or mission (ISO/IEC 15408-1, 2009; ISO/IEC 15408-2, 2008; ISO/IEC 15408-3, 2008).

Security is concerned with the *protection* of assets. The term protection, in the context of systems security engineering, has a very broad scope and is primarily oriented on the concept of assets and asset loss. Thus, the protection capability provided by a system goes beyond prevention and has the objective to control the events, conditions, and consequences that constitute asset loss. This includes the protection of intellectual property in the form of data, information, methods, techniques, and technology that are used to create the system or that are incorporated into the system. Thus, security is an *emergent property* of a system. Emergent properties are typically qualitative in nature, are subjective rather than objective in their nature and assessment and require consensus agreement based on evidentiary analysis and reasoning (NIST, 2018). *Emergent behaviour* is a design objective of all systems, but the emergent performance of a system can be desirable or undesirable, that produces a loss of assets. It might be forced by an attack, or in cases of incorrect application or mishandling of the system.

*System security* is the application of engineering and management principles, concepts, criteria, and techniques to optimize security within the constraints of operational effectiveness, time, and cost throughout all stages of the system lifecycle. Systems security engineering focuses on the protection of stakeholder and system assets. This protection is achieved by execution the specific activities and tasks in the system lifecycle processes with the objective of reducing vulnerabilities and constraining the impact of exploiting or triggering those vulnerabilities. This approach helps to reduce the susceptibility of systems to a variety of hazards and threats including physical and cyber attacks, structural failures, natural disasters, and errors of omission and commission (NIST, 2018). Systems security engineering, as a specialty discipline of systems engineering, provides several distinct perspectives and focus areas which are beyond the scope of technology and engineering. Therefore, *Security Engineering* can be considered as an interdisciplinary field of knowledge.

To achieve security objectives all system security activities and considerations should be tightly integrated into the technical and managerial processes of an engineering effort. This means full integration of systems security engineering into general systems engineering, and not execution of system security as a separate set of activities. Thus, systems security engineering, as part of a multidisciplinary systems engineering can be characterized by the following tasks (based on NIST, 2018):

- Defining stakeholder security objectives, protection needs and concerns, security requirements, and associated validation methods and verification methods;
- Development of security perspective of the system architecture and design;

- Identification and evaluation of vulnerabilities and susceptibility to lifecycle hazards, and threats;
- Designing proactive and reactive security barriers and functions encompassed within a balanced strategy to control asset loss and associated loss consequences;
- Providing security considerations to control, information and communication systems with the objective to reduce errors, mistakes, faults, and weakness that may constitute security vulnerability;
- Perform profit and loss analysis of specific security systems and features to prepare of engineering trade-offs, and risk management decisions;
- Demonstrate evidence-based researches that security and trustworthiness claims for a system have been satisfied.

The different causes of asset loss can arise from a variety of conditions and events related to adversity. The basis of asset loss constitutes all forms of intentional, unintentional, accidental, incidental, misuse, abuse, error, weakness, defect, fault, and/or failure events and associated conditions. The relationship between events, conditions and unacceptable asset loss consequences has several forms, e.g. (NIST, 2018):

- Events and conditions for which there is established sufficient knowledge of their occurrence and the specific loss consequences that result;
- Events and conditions that might occur (e.g., expected, anticipated, forecasted, or possible) and which would result in unacceptable or currently undefinable consequences;
- Emergent events and conditions that can result from the dynamic behaviours, interactions, and outcomes among system elements, including the specific interactions among them producing a loss.

There is no system that can be engineered to be perfectly secure or trustworthy, and that is why protections are necessary in most systems. There are two forms of protection capability:

- Active protection – the security functions of the system that exhibit security protection behavior and therefore, have functional and performance attributes.
- Passive protection – the environment for the execution and construction of all security functions. Passive protection includes architecture, design, and the rules that govern behaviour, interaction, and utilization.

System security is optimized based on a balanced proactive and reactive loss prevention strategy. A proactive loss strategy includes planned measures that are engineered to address what can happen rather than what might happen. Proactive systems security engineering also includes planning for all possible failure, nevertheless, whether the failure results from adversarial or non-adversarial events. A reactive loss strategy assumes that despite the proactive planning of means and methods to protect assets against undesirable events and associated consequences, unanticipated and otherwise unforeseen adverse consequences will occur. The proactive and reactive strategies are often combined with each other, and balanced across all assets, stakeholders, and objectives, usually based on expert elicitations.

Unfortunately, there are no methods that can verify and validate the effectiveness of solutions designed to ensure full system security. There will never be absolute assurance, however, because of the inherent asymmetry in system security, it means that a system can be declared insecure by observation, but there is no observation that allows one to declare an arbitrary system secure (Herley, 2016). Thus, in summary, it can be said that any efforts to ensure an adequate level of security of systems should meet at least the following requirements:

- provide the security functions that provide system security capability;
- assure the security-driven constraints for all system functions; and
- protect data, information, technology, methods, and assets associated with the system throughout its lifecycle.

### 8.1.2  *The Lifecycle-based Systems Security Engineering Framework*

The concept of lifecycle-based System Security Engineering refers to all processes and activities associated with the system throughout its entire life, with the focus on specific security considerations. This concept is like the idea of operational continuity which is the basis of both Dependability and Resilience Engineering (Bukowski, 2019), and includes processes such as activities related to development; prototyping; analysis of alternatives; training; logistics; maintenance; sustainment; evolution; modernization; disposal; and refurbishment. The lifecycle-based concept of System Security Engineering may include, but is not limited to requests for information, demands for proposal, statements of work, source selections, development, and test environments, operating environments and supporting infrastructures, supply chains, logistics, maintenance, training, verification, and validation procedures (INCOSE, 2015).

Based on these assumptions, the systems security engineering framework was developed, which provides a conceptual view of the key contexts of the systems security engineering activities (McEvilley, 2015). The framework defines, bounds, and focuses the systems security engineering activities and tasks, both technical and nontechnical, towards the achievement of system's security objectives and presents a coherent, evidence-based case as to how those objectives can be achieved. The framework is general, which means that it is independent of system type and is not to be interpreted as a sequence of flows or process steps but rather as a holistic set of interacting contexts, checks, and balances. Based on this proposal, a modified framework was developed and is shown in Figure 8.1. The framework states three main contexts within which the systems security engineering activities are carried out, namely, the problem context, the solution context, and the trustworthiness context. Context one includes clarifying of the problem and consists of the following steps: defining objectives, requirements, measures, as well as discovery of concepts, procedures, and processes. The second context contains finding solutions to the given problem by defining security aspects, implementing best possible practices, and seeking evidence of effective and efficient solutions. The third context concerns the provision of acceptable trustworthiness by developing different case studies, demonstrating their positive results, and suggesting possible improvements. Establishing these contexts should ensure that the engineering of a system is driven by a comprehensive understanding of the problem articulated in a set of stakeholder security objectives which consider both, protection needs and security concerns. The three contexts of the systems security engineering framework share a common base for the holistic security research activities which employ concepts, principles, means, methods,

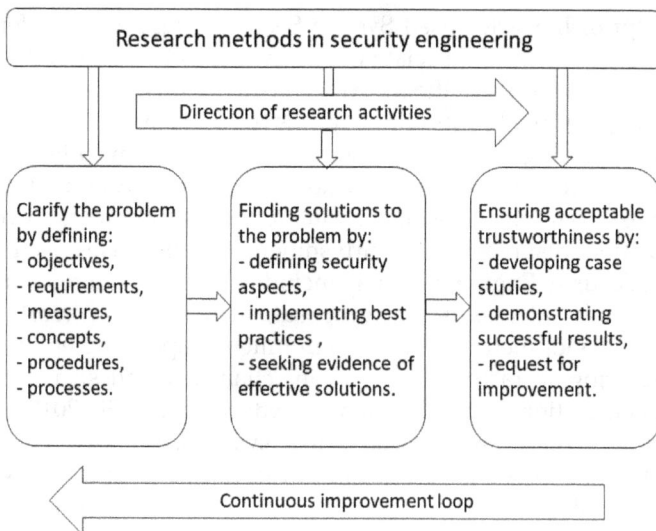

**Figure 8.1:** Process for improving the security level of a system [based on (McEvilley, 2015)].

processes, practices, tools, and techniques. Proper system security research can and should:

- Provide all relevant data and technical interpretations of system issues from the system security perspective;
- Ensure its application to align with the scope and objectives within the systems security engineering framework; and
- Secure a sufficient level of fidelity, rigour, and formality to produce data with a confidence that matches the requirements of the stakeholders and engineering team.

System security research must address all important topic areas related to systems security engineering including architecture; assurance; behaviour; cost; criticality; design; effectiveness; emergence; exposure; fit-for-purpose; lifecycle concepts; penetration resistance; performance; privacy; protection needs; requirements; risk; security objectives; strength of function; security performance; threat; trades; uncertainty; vulnerability; verification; and validation (NIST, 2018). The systems security engineering framework also includes a continuous improvement loop feedback for interactions among and between the three framework contexts. The feedback loop can also help to achieve continuous process improvement for the entire system.

## 8.2 Cyber Security – Principles, Concepts, and Practices

Cybersecurity is defined as: "The process of protecting information by preventing, detecting, and responding to attacks" (NIST, 2018). This protection is afforded to an automated information system to attain the applicable objectives of preserving the integrity, availability, and confidentiality of information system resources, includes hardware, software, firmware, information/data, and telecommunications. Thus, the three key objectives which are the foundation of cyber security must be clearly defined for the concept of security to be fully understood.

- Confidentiality covers two related concepts:
  - Data confidentiality – assures that private or confidential information is not available or disclosed to unauthorized individuals.
  - Privacy – ensures that individuals control or influence what information related to them may be collected and stored and by whom and to whom that information may be disclosed.
- Integrity consists of two elements:
  - Data integrity – assures that information and programmes are changed only in a specified and authorized manner.

o System integrity – ensures that a system performs its intended function in an unimpaired manner, free from deliberate or inadvertent unauthorized manipulation of the system.

- Availability – guarantees that systems work promptly, and service is not denied to authorized users only.

These three components of the cyber security concept are often referred to as the *CIA triad* and form the basis of security objectives for both data and for information and computing services. For example, the NIST standard FIPS 199 provides a useful in practice characterization of the CIA triad in terms of requirements and the definition of a loss of security in each category (FIPS, 2004):

- Confidentiality – preserving authorized restrictions on information access and disclosure, including means for protecting personal privacy and proprietary information. A loss of confidentiality is the unauthorized disclosure of information.

- Integrity – guarding against improper information modification or destruction, including ensuring information nonrepudiation and authenticity. A loss of integrity is the unauthorized modification or destruction of information.

- Availability – ensuring timely and reliable access to and use of information. A loss of availability is the disruption of access to or use of information or an information system.

The ever-increasing complexity of cyber systems and the growing demands on their security mean that, in some cases, meeting the CIA's triad requirements alone is not sufficient, and additional concepts are needed to present a complete systemic picture. In these cases, the following two criteria usually apply (FIPS, 2006):

- Authenticity – the property of being genuine and being able to be verified and trusted, confidence in the validity of a transmission, a message, or message originator. Verifying that users are who they say they are and that each input arriving at the system came from a trusted source.

- Accountability – the security goal that generates the requirement for actions of an entity to be traced uniquely to that entity. This supports nonrepudiation, deterrence, fault isolation, intrusion detection and prevention, and after-action recovery and legal action. Because fully secure systems are not yet an achievable goal, it must be possible to trace a security breach to a responsible party. Systems must keep records of their activities to permit later forensic analysis to trace security breaches or to aid in transaction disputes.

Federal Information Processing Standards (FIPS, 2004)) in Publication 199 uses three levels of impact on organizations or individuals of security, where there is a loss of confidentiality, integrity, or availability. These levels are defined as follows:

- Low – the loss could be expected to have a limited adverse effect on organizational operations, assets, or individuals. It means that the loss of confidentiality, integrity, or availability might: cause a degradation in mission capability to an extent and duration that the organization is able to perform its primary functions, but the effectiveness of the functions is noticeably reduced; result in minor damage to organizational assets; effect in minor financial loss or outcome in minor harm to individuals.

- Moderate – the loss could be expected to have a serious adverse effect on organizational operations, assets, or individuals. It means that the loss might: cause a significant degradation in mission capability to an extent and duration that the organization is able to perform its primary functions, but the effectiveness of the functions is significantly reduced; result in significant damage to organizational assets; effect in significant financial loss or outcome in significant harm to individuals that does not involve loss of life or serious, life-threatening injuries.

- High – the loss could be expected to have a severe or catastrophic adverse effect on organizational operations, assets, or individuals. A severe or catastrophic adverse effect means that the loss might: cause a severe degradation in or loss of mission capability to an extent and duration that the organization is not able to perform one or more of its primary functions; result in major damage to organizational assets; sustain major financial loss or outcome in severe or catastrophic harm to individuals involving loss of life or serious life-threatening injuries.

To unambiguously operationalize the basic concepts of cybersecurity assurance, it is necessary to define the basic terms necessary to build a security model. These are the following notions [based on (Stallings and Brown, 2018)]:

- Adversary (threat agent) – an entity that attacks, or is a threat to, a system.

- Attack – an assault on system security that derives from an intelligent threat, i.e., an intelligent act that is a deliberate attempt (especially in the sense of a method or technique) to evade security services and violate the security policy of a system.

- Countermeasure – an action, device, procedure, or technique that reduces a threat, a vulnerability, or an attack by eliminating or preventing it, by minimizing the harm it can cause, or by discovering and reporting it so that corrective action can be taken.
- Security risk – an expectation of loss expressed as the probability that a particular threat will exploit a particular vulnerability with a particular harmful result.
- Security policy – a set of rules and practices that specify or regulate how a system or organization provides security services to protect sensitive and critical system resources.
- System resource (Asset) – data contained in an information system, or a service provided by a system, or a system capability, such as processing power or communication bandwidth, or an item of system equipment (e.g., hardware), or a facility that houses system operations and equipment.
- Threat – a potential for violation of security, which exists when there is a circumstance, capability, action, or event, that could breach security and cause harm. In simple terms, a threat is a possible danger that might exploit a vulnerability.
- Vulnerability – a weakness in a system's design, implementation, or operation and management that could be exploited to violate the system's security policy. The following vulnerability categories are distinguished:
  - It can be corrupted, so that it does the wrong thing or gives wrong answers (e.g., stored data values may differ from what they should be because they have been improperly modified).
  - It can become leaky (e.g., someone who should not have access to the information available through the network obtains such access).
  - It can become unavailable or very slow. That is, using the system or network becomes impossible or impractical.

The relationship between these basic concepts is shown in Figure 8.2. It shows that the owner of the resource acts by creating an appropriate security policy, which is aimed at developing countermeasures to reduce the vulnerability of the resource, thereby reducing the level of risk. On the other hand, the adversary, called the threat agent, creates hazrds to the resources through various types of attacks, which in case of vulnerability to these threats (i.e., lack of proper security protections) create danger to the resources and assets.

The model shown in Figure 8.2 shows that an appropriate security policy can be an effective means to secure resources and assets. A

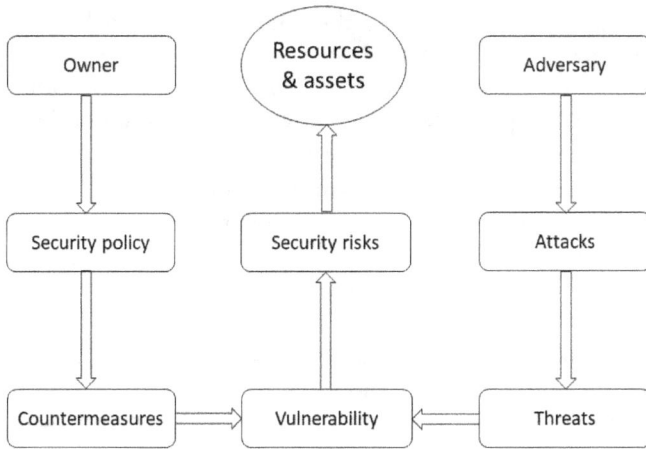

**Figure 8.2:** The relationship between basic security concepts.

comprehensive security policy involves three aspects: specification, implementation, and correctness assessment. The first step in devising security services and mechanisms is to prepare an informal description of desired system behaviour. Such informal policies may reference requirements for security, integrity, and availability. On this basis, a formal set of rules and practices is developed that specify or regulate how a system or organization provides security services to protect sensitive and critical system resources. In developing a security policy, the following factors, particularly, should be considered: the value of the assets being protected, all vulnerabilities of the system, and potential threats as well as the likelihood of possible attacks. Because security policy is also a business activity, influenced by legal requirements, decisions must be made with the trade-offs between:

- Ease of use versus highly secure. Practically all security measures involve some penalty in the ease of use. For example, firewalls and other efficient network security measures may reduce available transmission capacity or slow response time.

- Cost of increased security versus cost of failure and recovery. In addition to performance costs, there are direct financial costs in implementing and maintaining security measures. All these costs must be balanced against the cost of security failure and recovery if certain security measures are lacking.

The second step, security implementation, involves four complementary sequences of action:

- Prevention. An ideal security case is one in which no attack is successful, but there is a wide range of threats in which prevention is the only way to neutralize the effects of the attack. For example, by transmission of encrypted data, if a secure encryption algorithm is used, and if security barriers are in place to prevent unauthorized access to encryption keys, then attacks on confidentiality of the transmitted data will be prevented.

- Detection. In many cases, complete prevention of an attack is impossible, then it is practical to detect security threats. A practical example is detection of a denial-of-service attack, in which communications or processing resources are consumed so that they are unavailable for authorized users.

- Response. If security mechanisms detect an ongoing attack, such as a denial-of-service attack, the system should be immediately able to respond in such a way as to halt the attack and prevent further damage.

- Recovery. If prevention, detection, and response are not successful, recovery must be applied. An example of recovery is the use of backup systems – if data integrity is compromised, correct copy of the data should be reloaded.

The last, third step, concerns assurance and evaluation. *The NIST Computer Security Handbook* (NIST, 2018) defines assurance as the degree of confidence one has that the security measures, both technical and operational, work as intended to protect the system and the information it processes. This includes both system design and system implementation. Therefore, assurance understood as a degree of confidence deals with the questions: "Does the security system design meet its requirements?" and "Does the security system implementation meet its specifications?"

Evaluation is the process of examining a computer product or system with respect to given criteria. The central problem of work in this area is the development of evaluation criteria that can be applied to any security system (including security services and mechanisms).

In conclusion, Security Engineering is currently a very dynamically developing field of practical knowledge. Particular attention is paid to cyber systems, for which many standards and best practice recommendations have been developed. The following references give a selection of literature that is widely used and continuously improved to keep up with the development of modern technology (CNSSI, 2015; DHS Risk, 2010; DODD, 2015; FIPS, 2004; FIPS, 2006; ISO/IEC 15408-1, 2009; ISO/IEC 15408-2, 2008; ISO/IEC 15408-3, 2008; ISO/IEC 27001, 2013; ISO/IEC 27002, 2013; ISO/IEC 27034-1, 2011; ISO/IEC 27036-1, 2014; ISO/IEC 27036-2, 2014; ISO/IEC 27036-3, 2013; MIL-HDBK 1785, 1995; Moore, 2010; NASA, 2011; SP 800-137, 2011; SP 800-53A, 2014). On these foundations, Chapter 11 will present a comprehensive model for engineering of complex cyber-physical-social type systems.

## 8.3 References

Anderson, J. (1972). *Computer Security Technology Planning Study, Technical Report ESD-TR-73-51*. Air Force Electronic Systems Division, Hanscom AFB.

Bishop, M. (2005). *Introduction to Computer Security*, Boston, USA: Addison-Wesley.

Bukowski, L. (2019). *Reliable, Secure and Resilient Logistics Networks. Delivering Products in a Risky Environment*. Springer Nature Switzerland AG 2019, ISBN 978-3-030-00849-9 (Hardcover), ISBN 978-3-030-00850-5 (eBook); Doi: 10.1007/978-3-030-00850-5.

CNSSI. (2015). *Committee on National Security Systems Instruction (CNNSI) No. 4009*, Committee on National Security Systems Glossary, April.

DHS Risk. (2010). *Department of Homeland Security, DHS Risk Lexicon*.

DODD. (2015). *Department of Defense Directive 8140.01*. Cyberspace Workforce Management, August.

FIPS. (2004). National Institute of Standards and Technology, *Federal Information Processing Standards (FIPS) Publication 199*. Standards for Security Categorization of Federal Information and Information Systems, February. http://csrc.nist.gov/publications/fips/ fips199/FIPS-PUB-199-final.pdf .

FIPS. (2006). National Institute of Standards and Technology, *Federal Information Processing Standards (FIPS) Publication 200*. Minimum Security Requirements for Federal Information and Information Systems, March. http://csrc.nist.gov/publications/fips/fips200/FIPS-200-final-march.pdf.

Herley, C. (2016). *Unfalsifiability of Security Claims*. Microsoft Research, Proceedings of the National Academy of Sciences, 23 May.

INCOSE. (2015). *System Engineering Handbook—A Guide for System Engineering Life Cycle Processes and Activities*. International Council on Systems Engineering TP-2003-002-04, 4th Edn.

ISO/IEC 15408-1. (2009). International Organization for Standardization/International Electrotechnical Commission (ISO/IEC) 15408-1: 2009. *Information Technology — Security Techniques — Evaluation Criteria for IT Security — Part 1: Introduction and General Mod*el.

ISO/IEC 15408-2. (2008). International Organization for Standardization/International Electrotechnical Commission (ISO/IEC) 15408-2: 2008. *Information Technology — Security Techniques — Evaluation Criteria for IT Security — Part 2: Security Functional Requirements.*

ISO/IEC 15408-3. (2008). International Organization for Standardization/International Electrotechnical Commission (ISO/IEC) 15408-3: 2008. *Information Technology — Security Techniques — Evaluation Criteria for IT Security — Part 3: Security Assurance Requirements.*

ISO/IEC 27001. (2013). International Organization for Standardization/International Electrotechnical Commission (ISO/IEC) 27001: 2013. *Information Technology – Security Techniques – Information Security Management Systems – Requirements.*

ISO/IEC 27002. (2013). International Organization for Standardization/International Electrotechnical Commission (ISO/IEC) 27002: 2013. *Information Technology – Security Techniques – Code of Practice for Information Security Controls.*

ISO/IEC 27034-1. (2011). International Organization for Standardization/International Electrotechnical Commission (ISO/IEC) 27034-1: 2011. *Information Technology — Security Techniques — Application Security — Part 1: Overview and Concepts.*

ISO/IEC 27036-1. (2014). International Organization for Standardization/International Electrotechnical Commission (ISO/IEC) 27036-1: 2014. *Information Technology — Security Techniques — Information Security for Supplier Relationships — Part 1: Overview and Concepts.*

ISO/IEC 27036-2. (2014). International Organization for Standardization/International Electrotechnical Commission (ISO/IEC) 27036-2: 2014. *Information Technology — Security Techniques — Information Security for Supplier Relationships — Part 2: Requirements*.

ISO/IEC 27036-3. (2013). International Organization for Standardization/International Electrotechnical Commission (ISO/IEC) 27036-3: 2013. *Information Technology — Security Techniques — Information Security for Supplier Relationships — Part 3: Guidelines for Information and Communication Technology Supply Chain Security*.

Levin, T., Irvine, C., Benzel, T., Bhaskara, G., Clark, P. and Nguyen, T. (2007). *Design Principles and Guidelines for Security, Technical Report NPS-CS-07-014*. Naval Postgraduate School.

McEvilley, M. (2015). *Towards a Notional Framework for Systems Security Engineering*. The MITRE Corporation, NDIA 18th Annual Systems Engineering Conference, October.

MIL-HDBK 1785. (1995). Department of Defense Handbook. *System Security Engineering Program Management Requirements, MIL-HDBK-1785*.

Moore, J. (2010). *ISO/IEC/IEEE 15288 and ISO/IEC/IEEE 12207: The Entry Level Process Standards*. The MITRE Corporation.

NASA. (2011). *National Aeronautics and Space Administration System Safety Handbook*.

Neumann, P. (2004). *Principled Assuredly Trustworthy Composable Architectures. CDRL A001 Final Report*, SRI International, Menlo Park, CA, 28 December 2004.

NIST. (2018). *Framework for Improving Critical Infrastructure Cybersecurity*. National Institute of Standards and Technology. https://nvlpubs.nist.gov/nistpubs/cswp/nist.cswp.04162018.pdf.

SP 800-137. (2011). National Institute of Standards and Technology Special Publication (SP) 800-137. *Information Security Continuous Monitoring for Federal Information Systems and Organizations*. https://doi.org/10.6028/NIST.SP.800-137.

SP 800-53A. (2014). National Institute of Standards and Technology Special Publication (SP) 800-53A Revision 4. *Assessing Security and Privacy Controls in Federal Information Systems and Organizations: Building Effective Assessment Plans*. https://doi.org/10.6028/NIST.SP.800 -53Ar4.

Stallings, W. and Brown, L. (2018). *Computer Security: Principles and Practice*. UK: Pearson Education, Inc.

# 9

# Resilience Engineering
## The Concept of Process Continuity[*]

The topic of the chapter is the concept of resilience and its application in science and practice, especially in engineering. The foundations of a new approach to creating, implementing, and operating systems based on the so-called resilience thinking and its characteristics, including adaptability and capacity for transformation, are presented. A brief review of main development trends of the resilience approach in various areas of application was made and common features of these studies as fundamental for the emergence of a new discipline of knowledge – the Resilience Science – were summarized. Part one of the chapter concludes with a proposal for a matrix of strategies and aspects for providing resilience to systems and organizations. Part two of the chapter is devoted to a discussion of the principles, methods, and models used in resilience engineering. The basic relationships on which modelling the resilience of systems and organizations is based are presented, and five main types of models used in qualitative and quantitative research on resilience are analysed. At the end of the chapter, a universal simplified functional model of resilience is proposed, which will be a starting point for the development of complex quantitative-qualitative models in the following chapters of this work.

[*] https://orcid.org/0000-0002-2630-3507.

## 9.1 The Emergence and Development of Resilience Science

Resilience is a concept established in many academic fields and disciplines. *Resilience thinking* arose from observations of the reactions of living organisms to hostile environmental influences and has developed into an approach for understanding complex adaptive systems as a platform for multidisciplinary and transdisciplinary research with an emphasis on social-ecological systems (e.g., Levin et al., 2013). Social-ecological systems are composed of people and nature embedded in the biosphere (e.g., Folke, 2016).

The term resilience has been used in a narrow sense as the ability of a system to return to a state of equilibrium after the disturbance that knocked it out of that state has passed. In broad sense, the term resilience is used as an ability to bounce back after disturbance, or recovery. This can generally be interpreted as focus on trying to resist change and striving to maintain stability. The resilience approach deals with complex adaptive system dynamics and high uncertainty and seeks solutions how to live with change and even make use of it. In other words, resilience is having the capacity to persist in the face of change, and to continue to develop with ever changing environments. In practice, resilience research is about how periods of gradual changes interact with abrupt changes, and the capacity of people, communities, societies, cultures to adapt or even transform into new development pathways in the face of dynamic change. In resilience thinking, adaptation refers to human actions that sustain development on current ways, while transformation is about shifting development into other emergent scenario or even creating new ones (Folke, 2016). It is currently accepted that resilience reflects the ability of systems, people, communities, societies, and cultures to operate, live and develop with change, with ever-changing environments; cultivate the capacity to sustain development in the face of change, incremental and abrupt, expected, and surprising. Therefore, resilience is a dynamic concept concerned with managing complexity, uncertainty, and change across levels and scales on a human-dominated world.

*Adaptation* is a process of deliberate change in anticipation or in reaction to external stimuli and stress (Nelson et al., 2007). Adaptive capacity of people, communities, and societies are concepts in use in global environmental change in general and particularly, in climate change (e.g., Wise et al., 2014). The adaptability concept in resilience thinking captures the capacity of people in a social-ecological system to learn, combine experience and knowledge, innovate, and adjust responses and institutions to changing external drivers and internal processes. *Adaptability* has been defined as "the capacity of actors in a system to influence resilience" (Walker et al., 2004), and is central to persistence. It helps turn changes and surprises into opportunities and is an important part of social-ecological resilience (Berkes et al., 2003; Nelson et al., 2007).

*Transformability* is about shifting development into new pathways and creating novel ones, having the ability to cross thresholds and move systems into new basins of attractions, into new, emergent, and often unknown development trajectories (e.g., Walker et al., 2009; Marshall et al., 2012). Transformability has been defined as "the capacity to create a fundamentally new system when ecological, economic, or social structures make the existing system untenable" (Walker et al., 2004; Folke et al., 2010). There are several different ways of approaching transformations (e.g., Park et al., 2013). Transformative processes are characterized by discontinuities, thresholds, or tipping points and do not generally proceed smoothly. Thus, the resilience approach to transformations should also be about preparing for opportunity or creating conditions of opportunity for navigating the transformations (Chapin et al., 2009), and allowing the new identity of the social-ecological system to emerge through interactions of individuals, communities, and societies, and through their interplay with the biosphere within and across scales (e.g., Cumming et al., 2013; Folke et al., 2010). It concerns areas for safe-to-fail experimentation, facilitating different transformative experiments at small scales, and allowing cross-learning and new initiatives to emerge and spread across levels and scales (Westley et al., 2011; Biggs et al., 2015). Enhancing the resilience level of the new stability domain is a part of the transformation strategy (Chapin et al., 2009). Resilience, both for adaptability and transformability, operates and needs to be addressed across all levels and scales (Gunderson and Holling, 2002).

The *resilience approach* emphasizes that systems of humans and nature display nonlinear dynamics with thresholds, uncertainty, and surprise, and particularly, how such dynamics interact across temporal and spatial scales (e.g., Gunderson and Holling, 2002; Berkes et al., 2003). Complex systems have multiple attractors and there can be shifts from one attractor on a certain pathway to a new attractor and a contrasting pathway in a stability domain. The likelihood of such shifts increases with increasing vulnerability. Therefore, in resilience thinking, social and ecological systems are intertwined, exhibiting emergent properties and they can exist in qualitatively different states or basins of attraction. In complex adaptive systems, agents interact and connect with each other often in unpredictable and unexpected ways but from such an interaction broader scale patterns with new properties emerge, which feeds back on the system and influences the interactions of the agents. The resilience of individuals, groups, and communities is tightly coupled to this interplay and the emergent properties of the whole (Levin et al., 2013). Because complex adaptive systems can portray radically disproportional causation (i.e., small causes may produce big effects) and nonlinearity, they may depict periodic and chaotic dynamics, multiple basins of attraction, and potentially irreversible regime shifts (e.g., Biggs et al., 2009). Multiple drivers of change make it difficult to predict when such intense changes will occur and to pinpoint cause-and-effect mechanisms (e.g., Hughes et al., 2013). Living with such complexity and change requires adaptive approaches to decision making under high degrees of uncertainty and continuous learning as an important feature (Folke et al., 2005).

In the area of social-ecological systems, resilience is understood as the capacity of a specific system to continually self-organize and adapt in the face of ongoing change in a way that sustains the system in a certain stability domain or development directions. When analysing this capacity as a system property in relation to regime shifts it is useful to consider *resilience of what to what* (Carpenter et al., 2001). Biggs et al. (2015) defined resilience as the "capacity of a social-ecological system to sustain human well-being in the face of change, by persisting and adapting or transforming in response to given change". A central challenge in this context is the capacity of social-ecological systems to continue providing key ecosystem services that support human well-being in the face of unexpected shocks as well as gradual, ongoing change (Folke, 2016).

There is a search for objective measures of resilience which encounters various difficulties related primarily to moving targets continuously developing, evolving, and time-varying. Therefore, resilience as a system property should not be reduced to a one simple metric but consist of different types of metrics and indicators need to be used and combined to capture different facets of resilience. Several studies aim at developing resilience indicators in relation to regime shifts in diverse ecosystems, often with a focus on the interplay of fast and slow variables and feedback and how those are connected in dynamic landscapes and across scales, i.e., spatial resilience or with broader processes like rainfall patterns or fisheries and global seabird populations (Folke, 2016). Some researchers focus on measuring *resilience for who*m at the individual level, or on how to break resilience of one development path and transform it to another one, as well as to build *resilience of that path*. Whereas others concentrate on adaptation and adaptive capacity in relation to change with links to resilience thinking, there have been attempts to identify surrogates for resilience as well as sources of resilience (e.g., Goulden et al., 2013).

A critical concept for resilience management is *response diversity*, defined as the variety of responses to environmental change among species contributing to the same ecosystem function. Response diversity has been found to be particularly important for renewal and reorganization in ecosystems following change (e.g., Jansson and Polasky, 2010). However, a significant problem is the tension between the degree of simplification that resilience metrics demand and systems complexity. Resilience assessments aim at a deep understanding of system dynamics, recognizing that resilience is a dynamic property formed by many different processes of interacting fast and slow variables, including the larger context and cross-scale dynamics in which the social-ecological system is embedded as well as the unintentional changes of unforeseen dynamics (e.g., Quinlan et al., 2015).

Globalization in the modern world promotes the formation of relationships and interactions of various elements of social-ecological systems, and these relationships are dynamic. Cities both shape and are dependent on huge areas across the planet of ecosystems support for water, food, and other services (e.g., Folke, 2016) to create incentives for stewardship of their supporting ecosystems. The stewardship challenge is one of central focus in resilience thinking from early work on adaptive management to regime shifts and adaptive governance of social-ecological systems and ecosystem services (e.g., Folke, 2003). Therefore, ecological knowledge and understanding of ecosystem processes, and the social-ecological interplay of such dynamics is a prerequisite in this context. A dynamic interplay of actors, social networks, organizations, and diverse institutions, continuously learning with change, becomes significant features of social-ecological system dynamics, often emerging in relation to crisis situations as well as the opportunity for change toward stewardship of ecosystem services, and may lead to increased resilience on a higher governance scale.

There are also works on resilience in relation to legal structures, principles, and processes (e.g., Garmestani et al., 2013), core concepts of the rule of law (e.g., Ebbesson, 2010), as well as new forms of social contracts, emphasizing the dynamics, links, and complexity of social-ecological systems.

Scenario planning is a forward-looking approach aimed at predicting multiple alternative futures using qualitative and quantitative methods and data (e.g., Carpenter et al., 2009). In a complex and intertwined world, a surprise (e.g., an unexpected event) is to be expected (Kates and Clark, 1996). Resilience thinking, sometimes characterized as the science of surprise, is strongly shaped by underlying metaphors, speculations, and belief systems (Holling, 1986). Surprises involve not only extreme events, but also slow trends and weakly visible dynamic changes, so the resilience approach is very useful in these cases. In addition, this approach allows for the consideration of various types of uncertainty, including those that are not expressible in probabilistic terms (e.g., unknown unknowns). Therefore, *managing for resilience* enhances the likelihood of sustaining development in a rapidly changing world where unexpected should be expected (e.g., Chapin et al., 2009). When transformation is inevitable, resilient social-ecological systems contain the components needed for regeneration and reorganization, as well as reconnecting development for human well-being and sustainability (e.g., Biggs et al., 2015).

Specified resilience arises in response to the question "resilience of what to what?" and for "whom". Becoming too focused on specified resilience to increase resilience of parts or dimensions of a system to specific disturbances may cause the system to lose resilience in other critical disturbance situations. This means that there should be a trade-off between resilience of a social-ecological system to a small set of known kinds of disturbance versus the vast universe of unknown novel shocks. Specified resilience approaches may be narrowing options for dealing with novel surprises and even increasing the likelihood of new kinds of instability (Carpenter et al., 2015), thus, systems that become very robust to frequent kinds of disturbance necessarily become fragile in relation to infrequent ones (e.g., Folke, 2016).

Summarizing resilience science is an integrative approach for cultivating the capacity to sustain development in the face of change, incremental and abrupt, expected, and surprising, in relation to diverse pathways and thresholds and tipping points between them (Folke, 2016). From a transdisciplinary perspective, we can follow the authors (Ungar and Liebenberg, 2011) and describe resilience as a concept composed of seven elements:

- Access to material resources (availability of financial, educational, medical and employment assistance, resources, or opportunities, as well as access to food, clothing, and shelter).
- Relationships (relations with significant others, peers and adults within one's family and community).
- Identity (personal and collective senses of purpose, self-appraisal of strengths and weaknesses, aspirations, beliefs, and values, including spiritual and religious identification).
- Power and control (experiences of caring for one self and others; ability to affect change in one's social and physical environment to access health resources).
- Cultural adherence (adherence to one's local and global cultural practices, values, and beliefs).
- Social justice (experiences related to finding a meaningful role in community and social equality).

- Cohesion (balancing one's personal interests with a sense of responsibility to the greater good; feeling of being a part of something larger than oneself, socially and spiritually).

The above elements can be used to create strategies for ensuring the resilience of systems and organizations. Masten (2015) proposes three main directions for such strategies, namely:

- Risk-focused (prevent and reduce risk or adversity exposure).
- Asset-focused (increase resources or access to resources).
- Process-focused (restore or harness the power of human adaptive systems).

Table 9.1 attempts to combine these concepts into a one coherent whole, whereby the optimal strategy for a particular system or organization may be a combination of all these individual strategies.

Table 9.1: Strategies and aspects of providing resilience to systems and organizations.

| Strategy Aspects | Risk-focused | Asset-focused | Process-focused |
|---|---|---|---|
| Access to resources | availability | resources quality | supply continuity |
| Relationships | risk-sharing | assets-sharing | information-sharing |
| Identity | risk awareness | asset level monitoring | community of goals |
| Power and control | risk monitoring | managing resource | process supervision |
| Cultural adherence | risk analysis | cultural values | cultural practices |
| Social justice | risk acceptance | community values | social processes |
| Cohesion | risk reduction | resource balancing | liability trade-off |

Assuming, following Folke (2016), that the foundations of resilience science were laid in the early 1970s by Holling (1973, 1986), one can conclude from the perspective of five decades of development of this relatively new discipline of knowledge that its further development will proceed in two major directions. The first one may be called *General resilience* and represents a more broad-spectrum type of resilience for building capacity of systems to adapt or transform in response to the unknown. It applies to all kinds of changes and shocks, including extreme, novel, and unexpected ones (Hall and Lamont, 2013). The most significant features that characterize general resilience are diversity, modularity, openness, reserves, feedbacks, nonlinearity, leadership, and trust (Carpenter et al., 2012), giving systems and organizations the ability to deal with ongoing gradual change, with deep uncertainty and surprise. Based on empirical work and case studies, Folke et al. (2003) proposed four key features of *general resilience-building* for adaptive capacity, features that interact across temporal and spatial scales as well as build the system's ability to cope with dynamics in social-ecological systems:

- learning to live with change and uncertainty;
- nurturing diversity for reorganization and renewal;
- combining different types of knowledge for learning; and
- creating opportunity for self-organization toward social-ecological sustainability.

These principles should be approached as processes for generating conditions that allow for resolving collective-action challenges associated with multiple trade-offs in complex social-ecological systems. They support reflection, learning, and adaptation in search of deep understanding of complex, multivariable, nonlinear, cross-scale, and changing social-ecological systems provide guidance on how to use it for governance of social-ecological systems and ecosystem services (Biggs et al., 2015; Clarvis et al., 2015).

The second trend emerged at the beginning of the 21st century due to the interaction of the dynamically increasing complexity of cyber-physical-social systems and the concomitant increase in the intensity and scale of threats, both natural (e.g., hurricane Katrina) and caused by hostile human actions (e.g., the attack on the World Trade Center attack). This is especially relevant for systems of high complexity and stretch such as critical infrastructures, global supply networks, and has become the domain of Resilience Engineering, as will be discussed in Section 9.2.

## 9.2 Resilience Engineering – Principles, Concepts, and Models

Modern complex systems (e.g., from ultra-large critical infrastructure, through electric, energy and telecommunication grids, to small smart devices), must consider an ever- increasing variability as well as increasingly competing goals. Such goals include effectiveness, efficiency, user attractiveness but also sustainability, safety, and security. Resilience Engineering can significantly contribute to improving safety and security as well as the adaptive capabilities of complex socio-technical systems when they face changes, and potentially disruptive events. It means preserving critical functionality and enabling fast recovery of complex systems when the systems have problems with unexpected disruptions or untypical events. Therefore, Resilience Engineering is an innovative approach to improving the resilience of systems with the help of the technical means and engineering sciences (Hollnagel et al., 2006; Hollnagel,

2009). The resilience of such systems can be defined as their capability to successfully maintain the following abilities (NAS, 2012):

- Phase I – Plan and prepare. Development of foundations to keep services available and assets functioning during a disruptive event,
- Phase II – Absorb. Sustain most critical asset functions and service availability while preventing or isolating the disruption,
- Phase III – Recover. Restore all functions and services availability to their pre-disruption functionality,
- Phase IV – Adapt. Using knowledge gained from the event, alter protocol, configuration of the system, personnel training, or other aspects to become more resilient in the future.

Resource Guide on Resilience (IRGC, 2016) proposes the technical science-driven Resilience Engineering approach based on 12 main resilience engineering objectives, namely:

- to provide engineering, technical, and natural science founded approaches and processes to achieve resilience of complex systems,
- to develop tailorable validated methods to conceptualize, design, develop, and assess resilient systems. Extension and where appropriate replacement of (classical) risk approaches,
- to extend or replace classical notions of risk analysis and management with resilience concepts, approaches, and methods.
- to allow for extended and novel perspectives on risk events, risk propagation, risk assessment, and risk control.
- to be better able to prepare for less expected, seldom, unexpected, unknown, or even unexampled – so-called "black swan" – events,
- to seamlessly link to and to extend classical notions of reliability engineering designed to handle minor statistical and systematic failures,
- to extend maintenance concepts to response and recovery approaches post major disruptive events,
- to use and improve on resilience indicators that are also relevant for the daily successes of systems and vice versa,
- to integrate physical security, safety, IT-security approaches using methods from all corresponding disciplines and aggregating their analyses semi-quantitatively,
- to provide at least conceptual technical and engineering processes and methods that are independent of the type of adverse or disruptive event considered,

- to achieve a high level of individual, organizational, and societal commitment of all actors: third party, decision-makers, developers, system designers, system assessment personnel, by including highly unlikely events and their long-term consequences within risk control,
- to ask for the input and feedback of end-users and third parties by asking for the level of local controllability of scenarios and the duration of scenarios as well as considering the rare risks.

A variety of approaches and methods have been developed to quantify resilience of which four main groups are noteworthy (Häring, 2009):

- Analytical resilience quantification, which is based on the combination of semi-quantitative resilience dimensional assessments, e.g., using as outer cycle the resilience management cycle or risk management cycle. A possible starting point is to focus on resilience objectives for each resilience management Phase I to Phase IV (NAS, 2012).

- Resilience expansions with respect to resilience dimensions, e.g., number of events, resilience phases affected, etc. For instance, the assessment may focus on the immediate response in case of multiple physical cyber events.

- Resilience trajectory propagation, mainly for using or combining probabilistic-statistical and standardized engineering-simulative approaches. This approach focuses on the consideration of multiple possible events and their forward and backward propagation.

- Based on socio-technical and cyber-physical system simulations. For example, recovery times of airport checkpoints after security-induced disruptions can be determined from simulations (Renger et al., 2015).

Such Resilience Engineering approaches can be useful for formulating overall resilience optimization objectives, namely to (Häring et al., 2016):

- optimize the probability of an acceptable level of total resilience of a system,
- minimize the probability of non-acceptable overall resilience level of a system,
- optimize the total chance for fulfilling resilience objectives,
- minimize the risks on resilience objectives.

To be able to optimize resilience, it is necessary to build appropriate models that consider the most important factors affecting the level of resilience. Typical models recommended by Gibson and Tarrant

(2010) will be presented below. These models were built on six basic assumptions:

- Resilience is an outcome. Resilience is not a process, management system, strategy, or predictive measurement, but a feature that can be observed following, and in response to a substantial change in circumstances.

- Resilience is dynamic. There is no metric or score that will describe resilience as a fixed, constant characteristic, but will change in response to volatility in the external environment and internal system or organization over time, it will increase or decrease as the context changes.

- Resilience is not a composite feature. It arises from a complex interplay of many factors, therefore, as circumstances change, the presence, importance, and contribution of each of these factors to resilience will change.

- Resilience is multidimensional. There is currently no single model that describes resilience, all existing models have some limitations. The more sophisticated models describe many aspects of resilience from complementary viewpoints, but not all.

- Resilience exists over a range of conditions from low resilience (vulnerable) to high resilience (resilient). Such a spectrum of resilience can be observed amongst different systems and organizations facing the same event. The level of resilience of an organization depends on its maturity and should be improved from a low end highly reactive state (e.g., a simple emergency response such as an evacuation), through proactive preparedness (e.g., having in place incident response and business continuity capabilities), and finally to achieving a state where it is adaptive to conditions of high uncertainty.

- Resilience is founded upon good risk management. Many experts believe that RE has a close relationship with risk management and the objectives of the two systems should be closely correlated.

A. The Integrated Functions Model of Resilience

Early models of organizational resilience were based on the concept of business continuity management (BCM). Over time, this concept has been shown to fail in "black swan" type cases (Taleb, 2010), practical examples of which have occurred with increasing frequency in the 21st century (for example as occurred in the Enron Collapse, Katrina hurricane, and the global financial crisis). The result of this experience was, among other things, the concept of an integrated functional model, the basis of which remained traditional risk management, but closely linked to concepts such as emergency management (EM), business continuity management (BCM), security and crisis management (Figure 9.1). Risk management provides a common understanding of how uncertainty arising from highly volatile environments can affect the organization's objectives and offers how these specialized capabilities can then address that problem. Resilience, according to this model, was a characteristic that emerged from the harmonious interaction of all these five management components.

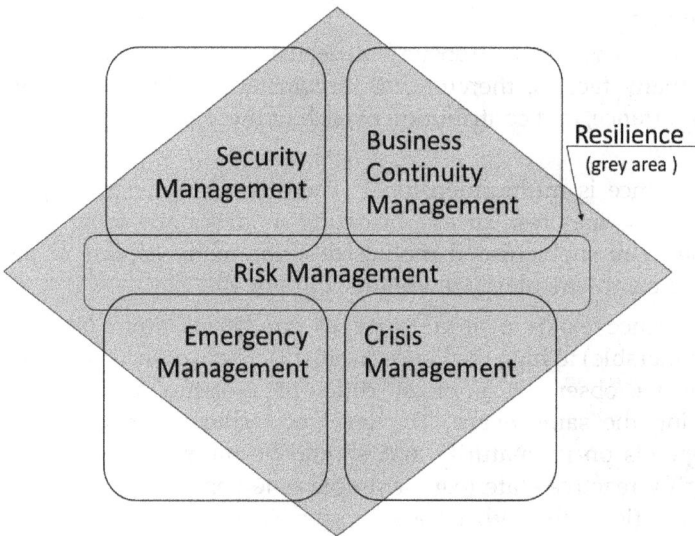

**Figure 9.1:** The integrated functions model of resilience.

B. Attributional Resilience Model

This model is the result of the work of the Resilience Community of Interest group and consists of the following two core elements, identified as key "drivers for creating resilience" (RCOI, 2009):

- The *organizational values* – establishing commitment, trust, and strong internal orientation for creating a common purpose.

- *Leadership* – creating a clear strategic direction based upon an understanding of systemic risks, empowering others to implement the strategic mission and vision, as well as stimulating trust.

These two main attributes are the foundation for creating an organizational culture and capability that is aware of, understands, and is sensitive to internal and external change and threats. The high level of *change sensitivity*, based on the past understanding, monitoring the present and predicting the future, allows indicators to be identified in preparing for unforeseen situations. The effective and efficient operation of the system is enabled through an Ren, adequate, and honest *communication process* that both provides an understanding and creates an *awareness* of how risks and threats to the organization are emerging or changing. It increases the organization's ability to learn from previous disruptions and allows better and understand and faster adapt to new emerging situations and conditions.

## C. Composite Resilience Model

The composite resilience model provides a different perspective that considers both soft and hard elements' operation: processes, infrastructure, technology, resources, information and knowledge. The main emphasis in this model is on the dominant role of long-term strategic activity, to which resource policies and management methods are subordinated. This allows for establishing an operational duality, the capability to operate in both routine and non-routine environments. It also means that emergent leadership does not necessarily arise only from top management, but often comes from talented and ambitious middle managers that rise to the occasion.

## D. Fishbone Model of Resilience

Based on the Ishikawa diagram, which was developed to graphically illustrate cause-and-effect relationships in quality management, a proposal was developed to use this type of model to analyse the resilience of complex organizations. Using this simple diagram, it is possible to represent the complex relationships within an organization that can determine its level of resilience to unforeseen changes and disruptions. Gibson and Tarrant (2010) list the following five critical factors that influence an organization's resilience performance:

- Acuity – the ability to recognize precedence (what has occurred in the past); situational awareness (what is happening now); and foresight (what could happen in the future) to identify early warning indicators of unexpected treats and provides an understanding of possible options for dealing with it.

- Ambiguity tolerance – the ability to making rational decisions and acting at times of high uncertainty.

- Creativity and agility – operating in novel ways to work around problems "unknown-unknown" type at a speed that matches volatility.

- Stress coping – supporting people, processes, and infrastructure continue to operate under increasing demands and uncertainty.
- Learnability – the ability to use the lessons of their own and others' experiences to better manage the prevailing conditions, including using information in real time as they emerge.

The relative contribution and importance to resilience of each of the capabilities, activities, and characteristics will depend upon the nature of the changing situations being faced by the organization.

E.  The Generic System Functions Model and its Modifications

The Generic System Functions Model has been proposed in different variants by many researchers, and the common feature of these alternatives is the triangular shape of the performance time-curve, so-called *resilience-triangle* (Sheffi, 2007; Abdelzaher and Kott, 2013). Sheffi and Rice (2005) defined eight phases for disruptive events: A – preparation, B – occurrence of a disruptive event, C – first response, D – initial impact, E – time of full impact, F – preparation for recovery, G – recovery, and H – long-term impact. The resilience depends on how big a decrease of the system performance was during these phases, and how much time is needed from the first impact of the disruptive event to the full recovery. The preparation phase is the time in which the system can anticipate and prepare for the disruptive event. This phase is very important for increasing a system's ability to bounce-back from a disruption.

An extended version of this model was proposed by Heinimann (2016) known as AR⁶A framework. The framework is based on eight generic system functions: attentiveness, robustness, resistance, restabilization, rebuilding, reconfiguration, remembering, and adaptiveness, which can be understood as following:

- *Attentiveness* and *remembering* are cognitive functions, capturing the ability to anticipate, and to detect endogenous and exogenous disruptions, and to memorize and learn from earlier disruptions, allowing a system to respond more rapidly and more effectively to future disruptions.
- *Robustness* refers to that range of performance (defined by an upper and lower limit) that guarantees a continually anticipated flow of services.
- *Resistance* is the critical limit of a system to withstand actions that are straining it during its lifecycle. If a disruption is damaging a system, its performance will decrease down to a minimum performance level, which can become zero in the worst case.

- The recovery phase consists of the two functions, *restabilize*, and *rebuild*, aiming to reestablish critical systems functions up to a range that enables survivability, to rebuild all the functions and to reestablish pre-disruption state (so-called "bouncing back" behaviour).

- *Reconfiguration* means to adapt and change systemic properties by introducing or deleting interdependencies, or some components. Experiences show that reconfiguration rarely happens with manmade systems but requires to change or enhance the system boundaries to adapt the topology of the system and make it more resistant and more resilient.

- *Adaptation* means to continue improvement of a system's abilities to cope with new threats, disturbances, and disruptions to increase its survivability.

Summarizing the above brief review of resilience models, it seems appropriate to adopt for further considerations in Chapter 11 the generalized model presented in Figure 9.2. This model is based on the basic assumptions of NAS (2012), and at the same time provides opportunities to develop it into an advanced quantitative model, which was attempted, for example, in the publication (Bukowski, 2016).

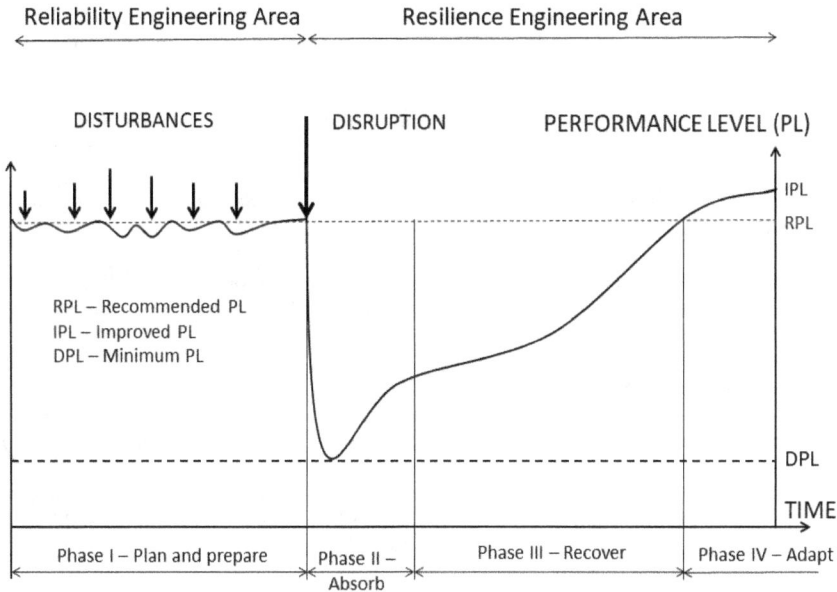

**Figure 9.2:** The generalized functional resilience model – a simplified version.

## 9.3 References

Abdelzaher, T. and Kott, A. (2013). Resiliency and robustness of complex systems and networks. *Adaptive, Dynamic and Resilient Systems*, 67: 67–86.

Berkes, F. (2007). Understanding uncertainty and reducing vulnerability: Lessons from resilience thinking. *Nat. Hazards*, 41: 283–295.

Berkes, F., Colding, J. and Folke, C. (eds.). (2003). *Navigating Social-ecological Systems: Building Resilience for Complexity and Change*. Cambridge, UK; Cambridge University Press. http://dx.doi.org/10.1017/cbo9780511541957.

Biggs, R., Carpenter, S.R. and Brock, W.A. (2009). Turning back from the brink: Detecting and impending regime shift in time to avert it. *Proceedings of the National Academy of Sciences of the USA*, 106: 826–831. http://dx.doi.org/10.1073/pnas.0811729106.

Biggs, R., Schlüter, M. and Schoon, M.L. (eds.). (2015). *Principles for Building Resilience: Sustaining Ecosystem Services in Social-ecological Systems*. Cambridge, UK: Cambridge University Press. http://dx.doi.org/10.1017/cbo9781316014240.

Bukowski, L. (2016). System of systems dependability–theoretical models and applications examples. *Reliability Engineering and System Safety*, 151: 76–92.

Carpenter, S., Walker, B., Anderies, J.M. and Abel, N. (2001). From metaphor to measurement: Resilience of What to What? *Ecosystems*, 4: 765–781.

Carpenter, S.R., Brock, W.A., Folke, C., van Nes, E. and Scheffer, M. (2015). Allowing variance may enlarge the safe operating space for exploited ecosystems. *Proceedings of the National Academy of Sciences of the USA*, 112: 14384–14389. http://dx.doi.org/10.1073/pnas. 1511804112.

Carpenter, S.R., Folke, C., Scheffer, M. and Westley, F. (2009). Resilience: Accounting for the noncomputable. *Ecology and Society*, 14(1): 13. [online] URL: http://www.ecologyandsociety.org/vol14/iss1/art13/.

Chapin, F.S., Kofinas, G.P. and Folke, C. (eds.). (2009). *Principles of Ecosystem Stewardship: Resilience-based Natural Resource Management in a Changing World*. New York, USA: Springer Verlag.

Clarvis, M.H., Bohensky, E. and Yarime, M. (2015). Can resilience thinking inform resilience investments? Learning from resilience principles for disaster risk reduction. *Sustainability*, 7: 9048–9066. http://dx.doi.org/10.3390/su7079048.

Cumming, G.S., Olsson, P., Chapin, F.S. and Holling, C.S. (2013). Resilience, experimentation, and scale mismatches in social-ecological landscapes. *Landscape Ecology*, 28: 1139–1150. http://dx.doi.org/10.1007/s10980-012-9725-4.

Ebbesson, J. (2010). The rule of law in governance of complex socioecological changes. *Global Environmental Change*, 20: 414–422. http://dx.doi.org/10.1016/j.gloenvcha.2009.10.009.

Folke, C. (2016). Resilience (Republished). *Ecology and Society*, 21(4): 44. https://doi.org/10.5751/ES-09088-210444.

Folke, C., Carpenter, S.R., Walker, B.H., Scheffer, M., Chapin III, F.S. and Rockström, J. (2010). Resilience thinking: Integrating resilience, adaptability, and transformability. *Ecology and Society*, 5(4): 20. [online] URL: http://www.ecologyandsociety.org/vol15/iss4/art20/.

Folke, C., Colding, J. and Berkes, F. (2003). Synthesis: Building resilience and adaptive capacity in social-ecological systems. pp. 352–387. *In*: Berkes, F., Colding, J. and Folke, C. (eds.). *Navigating Social-ecological Systems: Building Resilience for Complexity and Change*. Cambridge, UK: Cambridge University Press. http://dx.doi.org/10.1017/cbo9780511541957.020.

Folke, C., Hahn, T., Olsson, P. and Norberg, J. (2005). Adaptive governance of social-ecological systems. *Annual Review of Environment and Resources*, 30: 441–473. http://dx.doi.org/10.1146/annurev.energy.30.050504.144511.

Garmestani, A.S., Allen, C.R. and Benson, M.H. (2013). Can law foster social-ecological resilience? *Ecology and Society*, 18(2): 37. http://dx.doi.org/10.5751/ES-05927-80237.

Gibson, C. and Tarrant, M.E. (2010) A Conceptual models approach to organisational resilience. *Australian Journal of Emergency Management*, 25(2).

Goulden, M.C., Adger, W.N., Allison, E.H. and Conway, D. (2013). Limits to resilience from livelihood diversification and social capital in lake social-ecological systems. *Annals of the Association of American Geographers*, 103: 906–924. http://dx.doi.org/10.1080/00045608.2013.765771.

Gunderson, L.H. and Holling, C.S. (eds.). (2002). *Panarchy: Understanding Transformations in Human and Natural Systems*. Washington, D.C., USA: Island.

Hall, P.A. and Lamont, M. (eds.). (2013). *Social Resilience in the Neoliberal Era*. New York, USA: Cambridge University Press.

Häring, I., Ebenhöch, S. and Stolz, A. (2016). Quantifying resilience for resilience engineering of socio-technical systems. *European Journal for Security Research*, 1(1): 21–58. Doi: 10.1007/s41125-015-0001-x.

Häring, I., Schönherr, M. and Richter, C. (2009). Quantitative hazard and risk analysis for fragments of high-explosive shells in air. *Reliability Engineering & System Safety*, 94(9): 1461–1470. Doi: 10.1016/j.ress.2009.02.003.

Heinimann, H.R. (2016). *A Generic Framework for Resilience Assessment, in the IRGC Resource Guide on Resilience*, pp. 90–95, https://www.irgc.org/riskgovernance/resilience/.

Holling, C.S. (1973). Resilience and stability of ecological systems. *Annual Review of Ecology and Systematics*, 4: 1–23.

Holling, C.S. (1986). The resilience of terrestrial ecosystems: Local surprise and global change. pp. 292–317. *In*: Clark, W.C. and Munn, R.E. (eds.). *Sustainable Development of the Biosphere*. Cambridge, UK: Cambridge University Press.

Hollnagel, E. (2009). The four cornerstones of resilience engineering. pp. 117–133. *In*: Nemeth, C., Hollnagel, E. and Dekker, S. (eds.). *Resilience Engineering Perspectives: Preparation and Restoration*, Vol. 2. Burlington: Ashgate.

Hollnagel, E., Woods, D.W. and Leveson, N. (eds.). (2006). *Resilience Engineering: Concepts and Precepts*. Abingdon, Oxon, GBR: Ashgate.

Hughes, T.P., Linares, C., Dakos, V., van de Leemput, I.A. and van Nes, E.H. (2013). Living dangerously on borrowed time during slow, unrecognized regime shifts. *Trends in Ecology and Evolution*, 28: 149–155. http://dx.doi.org/10.1016/j.tree.2012.08.022.

IRGC. (2016). *Resource Guide on Resilience*. Lausanne: EPFL International Risk Governance Center. v29-07-2016 81. https://www.irgc.org/riskgovernance/resilience/.

Jansson, Å. and Polasky, S. (2010). Quantifying biodiversity for building resilience for food security in urban landscapes: Getting down to business. *Ecology and Society*, 15(3): 20. http://www.ecologyandsociety.org/vol15/iss3/art20/.

Kates, R.W. and Clark, W.C. (1996). Environmental surprise: Expecting the unexpected? *Environment*, 38: 6–34. http://dx.doi.org/10.1080/00139157.1996.9933458.

Levin, S., Xepapadeas, T., Crépin, A.-S., Norberg, J., de Zeeuw, A., Folke, C., Hughes, T.P., Arrow, K., Barrett, S., Daily, G., Ehrlich, P., Kautsky, N., Mäler, K.-G., Polasky, S., Troell, M., Vincent, J. and Walker, B.H. (2013). Social-ecological systems as complex adaptive systems: Modeling and policy implications? *Environment and Development Economics*, 18: 111–132. http://dx.doi.org/10.1017/S1355770X12000460.

Marshall, N.A., Park, S.E., Adger, W.N., Brown, K. and Howden, S.M. (2012). Transformational capacity and the influence of place and identity. *Environmental Research Letters*, 7: 034022. http://dx.doi.org/10.1088/1748-9326/7/3/034022.

Masten, A.S. (2015). Resilience in human development: Interdependent adaptive systems in theory and action. http://www.resilienceresearch.org/files/PTR/AnnMasten-PTRKeynote.pdf.

NAS. (2012). *Disaster Resilience: A National Imperative*. National Academy of Sciences (US).

Nelson, D.R., Adger, W.N. and Brown, K. (2007). Adaptation to environmental change: Contributions of a resilience framework. *Annual Review of Environment and Resources*, 32: 395–419. http://dx.doi.org/10.1146/annurev.energy.32.051807.090348.

Park, J., Seager, T.P., Rao, P.S.C., Convertono, M. and Linkov, I. (2013). Integrating risk and resilience approaches to catastrophe management in engineering systems. *Risk Analysis*, 33(3): 356–367.

Quinlan, A.E., Berbés-Blázquez, M., Haider, L.J. and Peterson, G.D. (2015). Measuring and assessing resilience: Broadening understanding through multiple disciplinary perspectives. *Journal of Applied Ecology*, 23: 677–687. http://dx.doi.org/10.1111/1365-664.12550.

RCOI. (2009). *Resilience COI, Report on the 2nd National Organisational Resilience Workshop 1–4 December 2008*. Published Sydney, February 2009.

Renger, P., Siebold, U., Kaufmann, R. and Häring, I. (2015). Semi-formal static and dynamic modeling and categorization of airport checkpoints. Nowakowski, T. Safety and Reliability: Methodology and Applications. *Proceedings of the European Safety and Reliability Conference, ESREL 2014*, Wrocław, Poland, 14–18 September 2014. Boca Raton, Fla.: CRC Press, 2015, pp. 1721–1731. Doi: 10.1201/b17399-234.

Sheffi, Y. (2007). *The Resilient Enterprise: Overcoming Vulnerability for Competitive Advantage*. Cambridge: MIT Press.

Sheffi, Y. and Rice, J.B. Jr. (2005). A supply chain view of the resilient enterprise. *MIT Sloan Management Review*, 47.1: 41–48.

Taleb, N.N. (2010). *The Black Swan: The impact of the Highly Improbable*. London: Penguin Books.

Ungar, M. (2016). The Resilience Research Centre Adult Resilience Measure (RRC-ARM); user's manual. http://www.resilienceresearch.org/files/CYRM/Adult%20-%20 CYRM%20 Manual.pdf.

Ungar, M. and Liebenberg, L. (2011). Assessing resilience across cultures using mixed methods: Construction of the child and youth resilience measure. *Journal of Mixed Methods Research*, 5(2): 126–149.

Walker, B., Holling, C.S., Carpenter, S.R. and Kinzig, A. (2004). Resilience, adaptability, and transformability in social-ecological systems. *Ecology and Society*, 9(2): 5. http://dx.doi.org/10.5751/es-00650-090205.

Walker, B.H., Abel, N., Anderies, J.M. and Ryan, P. (2009). Resilience, adaptability, and transformability in the Goulburn Broken Catchment, Australia. *Ecology and Society*, 14(1): 12. [online] URL: http://www.ecologyandsociety.org/vol14/iss1/art12/.

Westley, F., Olsson, P., Folke, C., Homer-Dixon, T., Vredenburg, H., Loorbach, D., Thompson, J., Nilsson, M., Lambin, F., Sendzimir, J., Banarjee, B., Galaz, V. and van der Leeuw, S.

(2011). Tipping towards sustainability: Emerging pathways of transformation. *Ambio,* 40: 762–780. http://dx.doi.org/10.1007/s13280-011-0186-9.

Westrum, R. (2006). A typology of resilience situations. pp. 49–60. *In*: Hollnagel, E., Woods, D.D. and Leveson, N. (eds.). *Resilience Engineering: Concepts and Precepts*, Aldershot, UK: Ashgate Press.

Wise, R.M., Fazey, I., Stafford Smith, M., Park, S.E., Eakin, H.C., Archer Van Garderen, E.R.M. and Campbell, B. (2014) Reconceptualising adaptation to climate change as part of pathways of change and response. *Global Environmental Change,* 28(2014): 325–336.

# 10

# Dependability Engineering
## The Concept of Fault-tolerant Functioning[*]

Chapter 10 provides the concept of Dependability Engineering as a relatively new, still a dynamically developing field of knowledge. A historical outline of the development of this concept is presented, starting with the idea of failure tolerance functioning and fault-tolerant systems. Based on these ideas, the Dependable Computing concept has matured and has been successfully applied to the field of computer science and information and communication technology. The second part of the chapter presents the basics of Reliability Engineering and the five selected approaches for building the foundation of this novel, emerging applied science discipline. Each of these approaches have its advantages and disadvantages, so their use should depend on the application area and the specifics of a given type of system. The vulnerability-based concept as well as risk related concept seem to be the most universal and therefore in the remainder of this work an attempt will be made to develop a universal model of Cognitive Dependability Engineering for cyber-physical-social systems on their assumptions.

## 10.1 Basic Concepts of Dependable Computing

The attribute-based Systems Engineering concepts discussed in Chapters 6, 7, 8, and 9 focused on only one dominant system attribute, namely, either

---

[*] https://orcid.org/0000-0002-2630-3507.

reliability, safety, security, or resilience. This was mainly a consequence of a reductionist approach to the complexity of systems and processes and a narrow specialization in the practice of building and operating real-world engineering systems. With the rapid development of applied computer science and computing techniques, this approach has proven insufficient. The main reason for this was the fact that the increasingly complex and sophisticated cyber-physical systems, consisting of hardware and software, in use were a single, forming an inseparable whole, item (or unit) whose correct operation depended equally on both these subsystems. The answer to these problems was the search for solutions focused not on one dominant feature of the system, but on the main operational goal of the entire system, which became a continuous, reliable, and safe operation, both under established operating conditions, i.e., "normal" and "abnormal".

Already in 1956, J. von Neumann, E.F. Moore, and C.E. Shannon developed theories of using redundancy to build reliable logic structures from less reliable components, whose faults were masked by the presence of multiple redundant components (Cristian, 1991). This theory was unified by W.H. Pierce (1965) as the concept of *failure tolerance*. Two years later, Avizienis (1967) integrated masking with the practical techniques of error detection, fault diagnosis, and recovery into the concept of *fault-tolerant systems*, and in 1975 work on software fault tolerance was done by B. Randell (1975). The emergence of a consistent set of new concepts and terminology resulted in the book *Dependability: Basic Concepts and Terminology* by J.C. Laprie in 1992 and a new research area *Dependable Computing* was born. The concepts are general enough to cover the entire range of computing and communication systems, from individual logic gates to networks of computers with human operators as well as users. The work focuses mainly on computing and communications systems, but the definitions are also intended in large part to be of relevance to computer-based systems, i.e., systems which also include the humans and organizations that provide the instantaneous environment of the computing and communication systems of interest.

*A system* in this taxonomy is understood as an entity that interacts with other entities, i.e., other systems, or subsystems, including hardware, software, humans, and the physical world with its natural phenomena, including risks, threats, and hazards. These other entities and systems create *the environment* of the given system, whereby the system boundary is the common frontier between the system and its environment. Computing and communication systems are characterized by five fundamental properties: *functionality, performance, dependability and security, and cost*. The function of such a system is what the system is intended to do (e.g., to operate, produce, or to deliver services) and is described by the functional specification in terms of functionality and performance. The behaviour of a system is what the system does to implement its function and is defined by a sequence of system's states. The *total state* of a given system is the set of the following states: computation, communication, stored information, interconnection, and physical condition.

*The structure of a system* (e.g., topology) is what enables it to generate the required behaviour. From a structural perspective, a system is composed of a set of components linked together to interact, where each component can be a simple item or a complex system. The recursion of the structure stops when a component is atomic, and any further internal structure is not of interest and can be ignored. Therefore, the comprehensive state of a system is the set of the states of its all-atomic components. *The service* delivered by a system as a provider is its behaviour as it is perceived by its user, and a user is another system that receives service from the provider. The part of the provider's system boundary where service delivery takes place is the provider's service interface. The delivered service is a sequence of the provider's external states. A system may be a provider and/or a user with respect to another system, i.e., deliver service to and receive service from another system. *The interface* of the user at which the user receives service is the use interface. A system can implement more than one function and deliver more than one service. Function and service can be thus seen as composed of function items and of service items.

*Correct service* is delivered when it implements the required system functions. A service failure is an event that occurs when the delivered service deviates significantly from the correct ones. A service failure period, so-called *a service outage*, is a transition from correct service to incorrect service and occurs either because it does not comply with the functional specification, or because this specification did not adequately describe the required system function. The transition from *incorrect service* to correct service is a service restoration. The deviation from correct service may assume different forms that are called service failure modes and are ranked according to failure severities. Since a service is a sequence of the system's external states, a service failure means that at least one external state of the system deviates from the correct service state. The deviation is called *an error*. The suspected or hypothesized cause of an error is called *a fault*. Faults can be internal or external of a system. The prior presence of *a vulnerability*, understood as an internal fault that enables an external fault to harm the system, is necessary for an external fault to cause an error and possibly subsequent failures. In most cases, a fault first causes an error in the service state of a component that is a part of the internal state of the system and the external state is not immediately affected. The specification may identify several such states, e.g., slow service, limited service, emergency service, e.g., in situations where the system is subjected to a partial failure of its functionality or performance.

Based on these assumptions the dependability of a system is defined as *the ability to deliver service that can justifiably be trusted*. This definition stresses the need for justification of trust. The criterion for justification of system dependability is based on the concept of risk and is understood as follows: *dependability of a system is the ability to avoid service failures that are more frequent and more severe than it is acceptable*. Generally, dependability is an integrative concept that consists of three parts: the attributes of, the means by, and the threats to which dependability is attained.

Dependability attributes:

- availability – readiness for correct service;
- reliability – continuity of correct service;
- safety – absence of catastrophic consequences on the users and the environment;
- confidentiality – absence of unauthorized disclosure of information;
- integrity – absence of improper system state alterations;
- maintainability – ability to undergo repairs and modifications.

Dependability means:

- fault prevention – how to prevent the occurrence or introduction of faults;
- fault tolerance – how to deliver correct service in the presence of faults (e.g., redundancy);
- fault removal – how to reduce the number or severity of faults (e.g., corrective or preventive maintenance);
- fault forecasting – how to estimate the present number, the future incidence, and the likely consequences of faults.

Dependability threats:

- faults;
- errors;
- failures.

Security is the concurrent existence of:

- availability for authorized users only;
- confidentiality;
- integrity with 'improper' meaning 'unauthorized'.

All potential faults that may affect a system during its life are classified according to eight basic viewpoints. These fault classes are called *elementary faults*. The classification criteria are as follows (Avizienis et al., 2004):

- The phase of system life during which the faults originate:
  - development faults that occur during (a) system development, (b) maintenance during the use phase, and (c) generation of procedures to operate or to maintain the system;
  - operational faults that occur during service delivery of the use phase.
- The location of the faults with respect to the system boundary:
  - internal faults that originate inside the system boundary;
  - external faults that originate outside the system boundary and propagate errors into the system by interaction or interference.
- The phenomenological cause of the faults:
  - natural faults that are caused by natural phenomena without human participation;
  - human-made faults that result from human actions.
- The dimension in which the faults originate:
  - hardware (physical) faults that originate in, or affect, hardware;
  - software (information) faults that affect software, i.e., programs or data.
- The objective of introducing the faults:
  - malicious faults that are introduced by a human with the malicious objective of causing harm to the system;
  - non-malicious faults that are introduced without a malicious objective.
- The intent of the human(s) who caused the faults:
  - deliberate faults that are the result of a harmful decision;
  - non-deliberate faults that are introduced without awareness.
- The capacity of the human(s) who introduced the faults:
  - accidental faults that are introduced inadvertently;
  - incompetence faults that result from lack of professional competence by the authorized human(s), or from inadequacy of the development organization.

- The temporal persistence of the faults:
  - permanent faults whose presence is assumed to be continuous in time;
  - transient faults whose presence is bounded in time.

If all combinations were possible, there would be 256 different combined fault classes. In practice, the number of most likely combinations is 31.

## 10.2 Fundamentals of Dependability Engineering

Over the years, the success of the dependability approach in computer science has contributed to the search for applications of this concept also in other areas of technology and engineering. As a result of these works there began to emerge, based on reliability engineering, its generalized version, namely *dependability engineering* (Kaâniche et al., 2000; IEC, 2015). A major advantage of this concept, especially for systems with high degree of complexity, was its service orientation and risk-based approach. The main reason for such a large interest in this area appears to be primarily due to the growing importance of the threats and hazards of the type so-called "LSLIRE" – Large Scale, Large Impact, Rare Events (e.g., Aven, 2015; Taleb, 2007; Taleb, 2012). Incidents of this type are particularly dangerous for a complex system such as critical infrastructure (e.g., El-Thalji, 2011; Kröger and Zio, 2011; Shafieezadeh and Burden, 2014; Zio, 2007) and global supply chains (e.g., Bosman, 2006; Kleindorfer and Saad, 2005; Natarajarathinam et al., 2009; Sheffi and Rice, 2005; Straube and Pfohl, 2009; Waters, 2007).

Therefore, it seems fully justified to strive for opportunities to generalize the concept of dependability in a way, that could be applied to all types of complex systems, especially cyber-physical systems, operating under both normal and abnormal work conditions. The first step towards solving this problem is to overview the basic concepts, principles, and models existing in this field and subsequently, there should be developed the general concept of dependability, applicable to all complex engineered systems (Bukowski, 2016, 2019).

a) Availability based concept of dependability – a probabilistic approach

The IEC Technical Committee 56 was formed in 1965 as 'Reliability and Maintainability' to prepare international standards that cover generic aspects of Reliability Engineering in the electro-technical field. In 1989, the initial title of IEC TC 56 was changed to 'Dependability', and in 1990, after consultations with quality related ISO 9000, the scope of the Committee was extended to the *generic dependability issues* across all technological disciplines. Based on these standards IEC 60050-191 and IEC 60300-1 (IEC, 2015) the term "dependability reflects user's confidence in fitness for use by attaining with satisfaction in product performance capability, delivering service availability upon demand, and minimizing the costs associated with the acquisition and the ownership throughout the life cycle". In this context, the product is understood to be both as a simple item (e.g., a material, a device, a machinery, an algorithm) and a complex system (e.g., transportation system, supply chain an integrated network comprising of hardware, software, and human elements).

In this view, dependability is defined as *the collective term used to describe the availability performance and its influencing factors: reliability performance, maintainability performance and maintenance support performance and it should be applied only for general, non-quantitative descriptions.* Availability performance is the ability of an item to be in a state to perform a required function under given conditions at a given instant of time or over a given time interval, assuming that the required external resources are provided. An item is defined as any part, component, device, subsystem, functional unit, equipment, or system that can be individually considered. The availability (AV) is defined as an outcome characteristic consisting of three elements:

- Reliability performance (REL) – the ability of an item to perform a required function under given conditions for a given time interval.
- Maintainability performance (MAI) – the ability of an item to be retained in or restored to a state in which it can perform a required function when maintenance is performed under given conditions and using stated procedures and resources.
- Maintenance support performance (MSP) – the ability of a maintenance organization to provide upon demand the resources required to maintain an item under given conditions and under a given maintenance policy.

A practical use of this concept has been presented in the standard IEC 60300-3-4 (Dependability management – Part 3: Application guide – Section 4: Guide to the specification of dependability requirements) (IEC, 2015). In the Annex A of this guide some examples of reliability, maintainability, maintenance support, and availability requirements are presented. As reliability performance measures are proposed:

- mean failure rate – $\lambda_m (t_1, t_2)$,
- mean time to failure – $MTTF$,
- mean failure intensity – $\zeta_m (t_1, t_2)$,
- mean time between failures – $MTBF$,

- useful life – $T_u$ and
- reliability function – $R\,(t_1, t_2)$.

The commonly used indicators for maintainability are usually mean repair time $MRT$ and mean time to restoration – $MTTR$. The typical measures for maintenance support are mean administrative delay – $MAD$ and mean logistic delay – $MLD$. The availability requirements are specified typically by mean availability – $A_m\,(t_1, t_2)$, mean unavailability $U_m\,(t_1, t_2)$, and mean down time – $MDT$. So, the specification of dependability requirements could be represented by the formula:

$$D = \{A(t_1, t_2); \lambda_m(t_1, t_2); R(t_1, t_2); MTBF; MTTF; MTTR; MRT; MAD; MLD; MDT\} \text{ (10.1)}$$

where the D-vector components are parameters of random variables distributions.

Estimating these parameters with statistical methods requires an adequate sample size, which is not always possible. This is particularly the case for systems characterized by a high level of reliability and of a low probability of hazards.

b) Availability and credibility related concept of dependability – a probabilistic-deterministic approach

The basis for this consideration is the concept of dependability presented in the standard IEC 61069: Industrial-process measurement and control – Evaluation of system properties for the purpose of system assessment, Part 5: Assessment of system dependability. In this standard, the term dependability is defined as: *the extent to which a system can be relied upon to perform exclusively and correctly a task under given conditions at a given instant of time or over a given time interval, assuming that the required external resources are provided. Dependability has two components of its property: availability as probabilistic component and credibility as deterministic ones.* The definition of availability is the same as in the IEC 60300-1 (IEC 2015), and its performance measures can be created in the same way, which means with support of statistical models and tools. *Credibility is defined as the extent to which a system is able to recognize and signal the state of the system and to withstand incorrect inputs or unauthorized access.* This deterministic property consists of the two components: *integrity* and *security*. Integrity is understood as the assurance provided by a system that the task will be performed correctly unless notice is given any state of the system, which could lead to the contrary. Security is the assurance provided by a system that any incorrect input, or unauthorized access is denied.

This concept was used by L. Bukowski and J. Feliks (2011) as the basis for the design of an expert model to assess the system dependability (D), which is characterized by availability (AV) and credibility (CR). In these works, the term credibility (CR) was understood as an amalgam of safety

(SAF) and security (SEC). There was proposed the following model for dependability measure:

$$D = \{AV \cap CR\}; \; AV = \{REL \cap MAI \cap MSP\}; \; CR = \{SAF \cap SEC\} \quad (10.2)$$

The authors used fuzzy logic procedure as well as linguistic variables to describe the expert knowledge about safety and security. The tool supporting the modelling and simulation of the expert system was WinFACT (2022), a specialized software. A base of rules for the subsystem CR with two inputs (SAF and SEC) and one output (CR), where every variable was divided into five linguistic categories (Very Low, Low, Moderate, High, and Very High), includes 25 elements. The correctness of selection of rules as well as the shape and ranges of the membership function is verified with rules viewer and simulation procedures (WinFACT User Guide). The rules viewer displays a roadmap of the entire fuzzy inference process. It also shows how the assumed shape of certain membership functions influences the overall result. Figure 10.1 presents a simulation model of a credibility evaluation system and the results of a simulation in the form of a graph showing the dependence of the output variable on the input variables. This model of the dependability deterministic part (credibility) can be integrated with the classical probabilistic part (availability) using fuzzification procedures.

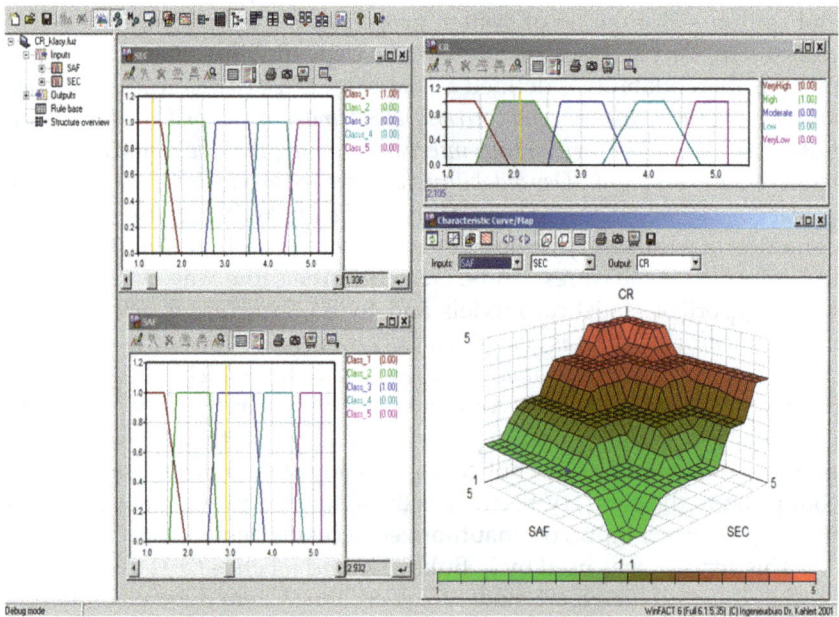

**Figure 10.1:** Credibility simulation model and its results (Bukowski and Feliks, 2011).

c) The concept of disruption tolerant networking

The Delay- and Disruption-Tolerant Network (DTN) is a network composed of smaller networks. It is a connection on top of special-purpose networks, including the Internet. DTN supports interoperability of other networks by accommodating long disruptions and delays between and within those networks, and by translating different communication protocols of those networks (Caini et al., 2011). The concept of DTN was originally developed for interplanetary use, where the delay-tolerance is extremely important. However, DTNs have diverse applications also on Earth, in situations where disruption-tolerance is the highest requirement. The potential of these applications extent a broad range of commercial, scientific, military, and public-service applications.

DTNs can use many kinds of wireless solutions, including radio frequency (RF), ultra-wide band (UWB), free-space optical (FSO), and acoustic (sonar or ultrasonic) technologies. Many evolving and possible communication environments are characterized by (Voyiatzis, 2012):

- Intermittent connectivity – the absence of an end-to-end path between source and destination (so-called network partitioning).

- Long or variable delay – long propagation delays between nodes and variable queuing time at nodes contribute to end-to-end path delays.

- Asymmetric data rates – the Internet supports moderate asymmetries of bidirectional data rate, but if asymmetries are large, they defeat conversational protocols.

- High error rates – bit errors on links require correction or retransmission of the entire packet.

DTN can overcome the problems associated with intermittent connectivity, long or variable delay, asymmetric data rates, and high error rates by using *store-and-forward message switching*. This idea is based on an old method, used by postal systems since ancient times. Whole messages (e.g., entire blocks of user data) or its fragments of such messages are moved forward from a storage place on one node (so-called switch intersection) to a storage place on another node, along a path to the given destination. The storage place (e.g., hard disk) can hold messages indefinitely, and they are called *persistent storage*, as opposed to very short-term storage provided for example by memory chips and buffers.

DTN routers need persistent storage for one of the following reasons (Caini et al., 2011):

- A communication link to the next 'hop' may not be available for a long time.

- One node in a communicating pair may send or receive data much faster or more reliably than the other node.
- A once transmitted message may need to be retransmitted if an error occurs at an upstream (toward the destination) node, or if an upstream node declines acceptance of a forwarded message.

DTN supports communication between intermittently connected nodes by isolating delay and disruptions with a store-and-forward technique. The intermittent connectivity may be opportunistic or scheduled.

The DTN store-and-forward message switching architecture is a generalization of work originally conceived to support the InterPlaNetary Internet (IPN). The primary goals are interoperability across network environments, and reliability capable of persisting hardware (e.g., networks) and software (e.g., protocols) failures. DTNs also may have several logistics applications, namely (Caini et al., 2011; Voyiatzis, 2012):

- cargo and vehicle tracking (by road, rail, sea, and air),
- autonomous vehicles communication and control,
- smart transportation networks,
- in-store and in-warehouse asset tracking,
- data transactions,
- processing-plant monitoring,
- infrastructure-integrity monitoring,
- disaster communication,
- atmospheric and oceanographic conditions,
- seismological events.

The concepts of *fault-tolerant computing* and *disruption-tolerant networking* became very successful in information technology and computer science, but was not general enough, to cover all types of cyber-physical systems, e.g., global production systems and logistics systems (e.g., supply chains and networks), especially in presence of the low frequent and unpredictable hazards.

d) Vulnerability related concept of dependability – a multidimensional approach

For cyber-physical systems, especially in situations where these systems operate under deep uncertainty and unpredictability, the approaches outlined in the previous sections (a through c) appear to be insufficient. To overcome these difficulties, a multidimensional approach based on the concepts of vulnerability and resilience is proposed. The concept of vulnerability is founded on three main fundamentals (Kroger and Zio, 2011), namely:

- Degree of loss and damages due to the impact of disruptive events;
- Degree of exposure to the risk sources, i.e., likelihood of being exposed to threats and hazards of a certain degree and susceptibility of an element at the risk of suffering loss and damages;
- Degree of resilience, i.e., the ability of a system to anticipate, cope with/absorb, resist, and recover from the impact of a threat (technical) or hazard (social).

In general, *vulnerability to a disruptive event* is understood as the degree to which a system is affected by a disruptive event and the term *dependability* in this view can be interpreted as *'anti-vulnerability'*. Therefore, within this concept the ability to provide of a service that can justifiably be trusted is called *dependability* and is used as a collective term describing the time-related operating quality of a system. This concept of dependability includes the constituent properties, that can be represented in the form of four main attributes (Bukowski 2019):

- *Availability (AV)* – ability to be in state to perform the required functions under given work conditions, is described by:
  - ○ *Reliability (REL)* – ability to perform the required functions, without failure, for a given time interval, under given work conditions;
  - ○ *Maintainability (MAI)* – ability to be retained in, or restored to a state to perform as required, under given conditions of use and maintenance;
  - ○ *Maintenance Support Performance (MSP)* – effectiveness of an organization in respect to maintenance support;
- *Safety (SA)* – ability to operate, normally or abnormally, without danger of causing human injury or death and without damage to the system's environment, it consists of:
  - ○ *Absence of Critical Damages (ACD)*;
  - ○ *Protection of the environment* against the effects of any potential critical damages *(PRO)*;
- *Security (SE)* – ability to prevent an unauthorized access to, or handling of system state, can be described by the concurrent existence of:
  - ○ *Confidentiality (CON)* – unavailability to non-enabled persons;
  - ○ *Integrity (INT)* – impossibility of introducing changes into the system by non-enabled persons;
  - ○ *Accessibility for enabled users only (ACC)*;

- *Resilience (RE)* – a collective term describing the ability of a system to absorb and withstand the failure impact, and still continue to operate at acceptable predefined performance level, is described by:
  - *Survivability (SUR)* – capability of a system to fulfil its function, in a timely manner, in the presence of failures (absorbability);
  - *Recoverability (REC)* – capacity of a system to recover from a failure, within the acceptable time and costs limits (restoration);
  - *Adaptability (ADA)* – ability to adapt to changed working conditions (flexibility, agility, ability to learn).

This dependability attributes and some typical examples of its metrics are shown in Table 10.1.

**Table 10.1:** Dependability attributes and its metrics.

| Attributes | Examples of Metrics |
|---|---|
| Availability | mean availability – $A_m$ $(t_1, t_2)$; mean down time – MDT; |
| Reliability | mean failure rate – $\lambda_m$ $(t_1, t_2)$; mean time to failure – MTTF; mean time between failures – MTBF; reliability function – R $(t_1, t_2)$ |
| Maintainability | mean repair time MRT; mean time to restoration – MTTR |
| Maintenance Support Performance | mean administrative delay – MAD; mean logistic delay – MLD |
| Safety | probability that a system will be fully functioning or will fail in a manner that causes no harm in the time-period $(t_1, t_2)$ |
| Security | probability that a protection subsystem of a main system will be able to prevent an unauthorized access to the system in the time-period $(t_1, t_2)$ |
| Resilience | probability that a system will be able to fulfil its function in the presence of failures in the time-period $(t_1, t_2)$; mean time to "bounce back" – MTTBB |

A general dependability model built on the above assumptions can be represented by the following equation:

$$D(t_1, t_2) = \{AV(t_1, t_2); SA(t_1, t_2); SE(t_1, t_2); RE(t_1, t_2)\} \tag{10.3}$$

In practice, this model can be interpreted as follows:

*A system's dependability is the collective term, that describes its ability to continuous, safe, and secure fulfilment the required functions in a risky environment.*

e) Risk related concept of dependability – a two-dimensional perspective

With reference to the definition of reliability presented in Section 10.1 (Avizienis et al., 2004), namely: "dependability of a system is the ability to avoid service failures that are more frequent and more severe than it is acceptable", a two-dimensional concept of dependability is proposed, based on the diagram shown in Figure 10.2. This diagram shows a simplified model of a cyber-physical system from two perspectives: the operational perspective, which focuses on ensuring the continuity of processes for the flow of resources, goods, and services (supply and delivery); and the risk perspective, which focuses on minimizing the negative effects of business continuity disruptions (impacts).

The general model of a cyber-physical system is defined by its structure ($E \times R$) and behaviour ($X \times Y$) in a 2-dimensional space. From the operational perspective the system input $X$ is a vector representing all controllable events (supply of resources), and the output $Y$ is a vector which represents the produced functions (delivery of goods and services). In the risk perspective however, the system inputs are the uncontrollable events occurring in its environment (risk sources), which can generate risky scenarios. To prevent the negative effects of these scenarios, the system is equipped with security barriers. If these prove ineffective, the system is compromised by exposure to a given threat, which can result in a disruptive event. A properly designed system should be equipped with safety barriers that are designed to minimize the negative effects of an event, resulting in certain disruption consequences. This model can provide a basis for analysing the dependability of a complex engineered system which operates in a risky environment, using the *risk related concept of dependability*.

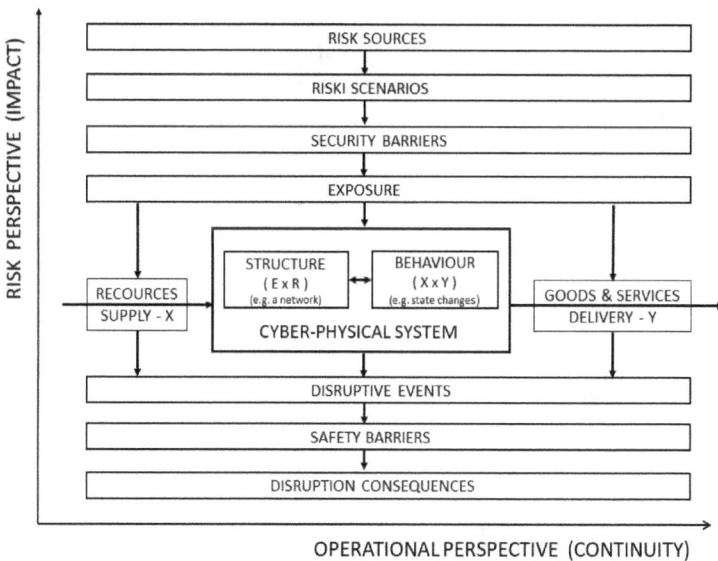

Figure 10.2: General model of a cyber-physical system in two-dimensional perspective.

Using this concept and based on the model shown in Figure 10.2, the following mathematical formula for quantifying dependability in a probabilistic approach is proposed:

$$D(t_1, t_2) = P\{[P(DE_m) < P_{cr}(DE)] \cap [V(DE_m) < C_{cr}(DE)]\} \qquad (10.4)$$

with:

$D(t_1, t_2)$ – dependability measure of a system in the time interval from $t_1$ to $t_2$

$P(DE_m)$ – the probability of a disruptive event $DE_m$ in the time interval from $t_1$ to $t_2$

$V(DE_m)$ – system's vulnerability to a disruptive event $DE_m$

$P_{cr}(DE)$ – critical value of disruption probability (frequency)

$C_{cr}(DE)$ – critical value of disruption consequences.

Also, the criterion for a cyber-physical system dependability justification based on the concept of risk can be described as follows:

DEPENDABILITY MEASURE OF A GIVEN CYBER-PHYSICAL SYSTEM IS THE LIKELIHOOD TO AVOID DISRUPTIONS THAT ARE MORE PROBABLE AND MORE SEVERE IN CONSEQUENCES THAN IT IS ACCEPTABLE FOR EACH POSSIBLE RISKY SCENARIO.

In conclusion, it can be said that Dependability Engineering is a relatively new, still a dynamically developing field of knowledge. This section presents five selected approaches for building the foundation of the novel, emerging applied science discipline. Each of these approaches have their advantages and disadvantages, so their use should depend on the application area and the specifics of a given type of system. Chapter 11 will attempt to develop a universal model of Cognitive Dependability Engineering, adapted to the requirements of complex cyber-physical-social systems.

## 10.3 References

Aven, T. (2015). Implications of black swans to the foundations and practice of risk assessment and management. *Reliability Engineering and System Safety*, 134: 83–91.

Avizienis, A. (1967). Design of fault-tolerant computers. *Proc. 1967 Fall Joint Computer Conf., AFIPS Conf. Proc.*, 31: 733–743.

Avizienis, A., Laprie, J.-C., Randell, B. and Landwehr, C. (2004). Basic concepts and taxonomy of dependable and secure computing. *IEEE Trans. on Dependable and Secure Computing*, 1(1): 11–33. https://www.nasa.gov/pdf/636745main_day_3-algirdas_avizienis.pdf.

Bosman, R. (2006). The New Supply Chain Challenge: Risk Management in a Global Economy, Johnston, RI: FM Global.

Bukowski, L. (2016). System of systems dependability - Theoretical models and applications examples. *Reliability Engineering & System Safety*, 151: 76–92, ISSN 0951-8320. Doi: 10.1016/j.ress.2015.10.014 0951-8320.

Bukowski, L. (2019). *Reliable, Secure and Resilient Logistics Networks. Delivering Products in a Risky Environment*. Springer Nature Switzerland AG 2019. ISBN 978-3-030-00849-9 (Hardcover), ISBN 978-3-030-00850-5 (eBook).

Bukowski, L. and Feliks, J. (2011). Evaluation of technical systems dependability with the use of fuzzy logic and experts' knowledge. *Proceedings of the 15th World Multi-Conference on Systemics, Cybernetics and Informatics, Orlando (Florida)*.

Caini, C., Cruickshank, H., Farrell, S. and Marchese, M. (2011). Delay- and Disruption-Tolerant Networking (DTN): An alternative solution for future satellite networking applications. *Proceedings of the IEEE*, 99(11): November 2011. https://www.researchgate.net/publication/224248901_Delay-_and_Disruption-Tolerant_Networking_DTN_An_Alternative_Solution_for_Future_Satellite_Networking_Applications.

Cristian, F. (1991). Understanding fault-tolerant distributed systems. *Comm. ACM*, 34(2): 56–78.

El-Thalji, I. (2011). System Dependability Engineering for Wind Power Application. http://www.diva-portal.org/smash/get/diva2:443903/fulltext01.

https://www.kuenzigbooks.com/pages/books/28503/e-f-moore-c-e-shannon-edward-claude-elwood/reliable-circuits-using-less-reliable-relays-first-collected-edition-the-only-known-signed-copy.

IEC (2015). IEC 6030 – Dependability Management – Part 3: Application Guide – Section 4: Guide to the Specification of Dependability Requirements.

IEC (2015a). International Standards on Dependability, The World's Online Electrotechnical Vocabulary, 2015. www.electropedia.org/.

Kaâniche, M., Laprie, J.C. and Blanquart, J.P. (2000). *Dependability Engineering of Complex Computing Systems*. 6th International Conference on Engineering of Complex Computer Systems (ICECCS 2000); 11–14 September 2000, Tokyo, Japan. http://homepages.laas.fr/kaaniche/Slides/Pres_ICECCS%2700.pdf.

Kleindorfer, P.R. and Saad, G.H. (2005). Managing disruption risks in supply chains. *Production and Operations Management*, 14(1): 53–68.

Kröger, W. and Zio, E. (2011).*Vulnerable Systems*. London: Springer-Verlag.

Laprie, J.C. (1992). *Dependability: Basic Concepts and Terminology*. London: Springer-Verlag.

Natarajarathinam, M. et al. (2009). Managing supply chains in times of crisis: A review of literature and insights. *International Journal of Physical Distribution and Logistics Management*, 39: 535–573.

Pierce, W.H. (1965). *Failure-tolerant Computer Design*. Paris: Academic Press.

Randell, B. (1975). System structure for software fault tolerance. *IEEE Trans. Software Eng.*, 1(2): 220–232.

Shafieezadeh, A. and Burden, L.I. (2014). Scenario-based resilience assessment framework for critical infrastructure systems: Case study for seismic resilience of seaports. *Reliability Engineering and System Safety*, 132: 207–219.

Sheffi, Y. and Rice, Jr. J.B. (2005). A supply chain view of the resilient enterprise. *MIT Sloan Management Rev.*, 47: 41–48.

Straube, F. and Pfohl, H.C. (2009). Global Networks in an Era of Change. *Environment, Security, Internationalization, People, Hamburg*.

Taleb, N.N. (2007). *The Black Swan: The Impact of the Highly Improbable*. London: Penguin.

Taleb, N.N. (2012). *Anti-fragile*. London: Penguin.

Voyiatzis, A.G. (2012). A survey of delay- and disruption-tolerant networking applications. *Journal of Internet Engineering*, 5(1): 331–344. http://www.artemiosv.info/hosted/ JIE-DTN-Survey.pdf.

Waters, D. (2007). *Supply Chain Risk Management: Vulnerability and Resilience in Logistics*. London & Philadelphia: Kogan Page Limited.

WinFACT. (2022). User Guide.

Zio, E. (2007). From complexity science to reliability efficiency: A new way of looking at complex network systems and critical infrastructures. *International Journal of Critical Infrastructure*, 3(3/4): 488–508.

# 11

# Cognitive Dependability Engineering

## The Concept of Trustworthy Performance*

This chapter is devoted to an extended version of Dependability Engineering, namely Cognitive Dependability Engineering (CDE). It deals with systems of very high complexity, called *human-centric cyber-physical-social systems,* and the research perspective of these systems is focused on the continuity of operation and functioning. At the core of the CDE is the concept of *trustworthy performance,* and the research method recommended for such defined cognitive objects is transdisciplinarity. The chapter firstly discusses key features of scientific theories, and conditions of scientific cognition according to the Einstein's model. Then the process of emergence of a new science, which can be divided into three main stages, is deliberated upon. According to this model, has been proposed the general concept of Cognitive Dependability Engineering and its area of applicability depending on the complexity of the systems and the level of uncertainty. The foundations of the elicitation procedure as a basis for cognitive dependability assessment of complex systems are presented, and a pre-paradigm for a new field of knowledge, Cognitive Dependability Engineering, is presented.

* https://orcid.org/0000-0002-2630-3507.

## 11.1 Defining Theoretical Assumptions of Scientific Theories

The word *'theory'* is ambiguous and has many meanings in everyday language. The most important of these are quoted in the Merriam-Webster Dictionary (https://www.merriam-webster.com/dictionary/theory) as follows:

- A plausible or scientifically acceptable general principle or body of principles offered to explain phenomena (the wave theory of light).
- A belief, policy, or procedure proposed or followed as the basis of action (her method is based on the theory that all children want to learn).
- An ideal or hypothetical set of facts, principles, or circumstances — often used in the phrase in theory (in theory, we have always advocated freedom for all).
- A hypothesis assumed for the sake of argument or investigation.
- An unproved assumption: conjecture.
- A body of theorems presenting a concise systematic view of a subject (theory of equations).
- The general or abstract principles of a body of fact, a science, or an art (music theory).
- Abstract thought: speculation.
- The analysis of a set of facts in their relation to one another.

Therefore, before we proceed to propose a procedure for building the theoretical foundations of cognitive reliability, let us try to define the concept of *scientific theory*. This is neither a simple nor an easy task since there are many different approaches to this problem in the scientific community. In the following we will try to present proposals representing the most common views, systematized by dividing them into three dominant groups.

The first group includes the 'metaphorical' approach, represented by Popper and Hooker. Popper's (1959) in his work *The Logic of Scientific Discovery* described the theory as:

"Theories are nets cast to catch what we call 'the world': to rationalize, to explain, and to master it. We endeavour to make the mesh ever finer and finer." (p. 59.)

Hooker (1987) proposed a metaphor of a pyramid, which more adequately reflected the essence of the concept of theory:

"At the bottom of the deductive pyramid lie the so-called observation sentences — those sentences whose truth can be checked experimentally—

whilst at the apex of the pyramid lay the most general theoretical principles of the scheme. Just exactly where the twin elements of theory and observation permeate this structure is a matter of contemporary controversy." (p. 109.)

Hooker's pyramid metaphor provides two important clues for developers of theories based on empirical research. First, that as many observational results as possible should form the basis for as few theoretical concepts as possible (information compression). The second contribution of Hooker's pyramid metaphor is the idea that theories are meant to be universal, synoptic statements, concerning a possible wide range of applications. Thus, theories can be seen as intellectual generalizations that maximize the information content of knowledge by reducing large sets of concrete observations to a much smaller number of explicit terms and concepts.

The second group of definitions has a more classical character, although there are quite significant differences in the way the problem is approached by different researchers. For example, Kaplan (1964) proposes a descriptive definition of theory as follows:

"Theory [is] the device for interpreting, criticizing, and unifying established laws, modifying them to fit data unanticipated in their formulation, and guiding the enterprise of discovering new and more powerful generalizations. To engage in theorizing means not just to learn by experience, but also to take thought about what there is to be learned." (p. 295.)

"A theory is a symbolic construction." (p. 296), and "Theory is thus contrasted with both practice and with fact." (p. 296.)

A much more compact and concise definition of the theory was proposed by Rudner (1966) in his work on the philosophy of social science:

"A theory is a systematically related set of statements, including some law-like generalizations, that is empirically testable." (p. 10.)

However, Kerlinger (1986) in a widely used introductory text defined theory in this manner:

"A theory is a set of interrelated constructs (concepts), definitions, and propositions that present a systematic view of phenomena by specifying relations among variables, with the purpose of explaining and predicting the phenomena." (p. 9.)

Another approach, used mainly in psychology, assumes that a theory is characterized by the ability to logically describe, by a process of formal deduction, the relations between a quantity of stimuli and responses. For instance, Hull (1943) wrote:

"A theory is a systematic deductive derivation of the secondary principles of observable phenomena from a relatively small number of primary principles or postulates." (p. 2.)

In contrast, the definition of Rose (1954) for social sciences is more specific and takes the following form:

"A theory may be defined as an integrated body of definitions, assumptions, and general propositions covering a given subject matter from which a comprehensive and consistent set of specific and testable hypotheses can be deduced logically. The hypotheses must take the form 'If a, then b, holding constant c, d, e . . .' or some equivalent of this form, and thus permit causal explanation and prediction." (p. 3.)

The third group of definitions focuses on describing the requirements that theories should fulfil to be considered formulated in accordance with scientific requirements. These include, for example Wacker's (1998) proposal:

- theory provides a framework for analysis;
- it offers an efficient method for field development; and
- it delivers clear explanations for the pragmatic world.

The first condition can be interpreted as a requirement that allows for a generalized model of the procedure for analysing the problems addressed by the theory. The efficiency of a theory is to ensure that the potential for error because of applying the theory is minimized. In practice, this means the need to integrate all available research results from a given field into a single, coherent body of knowledge. The third condition boils down to fulfilling the requirements of applicability of a given theory, i.e., its practical usefulness in the widest possible spectrum of problems and cases.

Wacker (1998) believes that meeting the above conditions is possible if the theory has a structure consisting of four basic components:

- definitions of terms or variables,
- a domain where the theory applies,
- a set of relationships of variables, and
- specific predictions factual claims.

to provide answers to the basic questions asked by practitioners, namely, who, what, when, where, how, why, should, could and would.

Rose (1954), for example, suggested, that "There are certain dangers in the use of theory in science" and then offered the following caveats concerning those dangers:

1. Theory channelizes research along certain lines.
2. Theory tends to bias observations.

3. The concepts that are necessary in theory tend to get reified.
4. ... [since] replications of a study seldom reach an identical conclusion ... are we justified in formulating elaborate theories which assume consistency in findings?
5. Until a theory can be completely verified, which is practically never, it tends to lead to over generalization of its specific conclusions to areas of behavior outside their scope.
6. ... rival theories of human behaviour ... seem[s] to encourage distortion of simple facts (pp. 4–5).

Based on his own experience as a long-time editor of a scientific journal, Feldman (2004) summarizes the most important characteristics of good theories as follows:

- the research problem must not be trivial in scope,
- the new theory should have a solid reference to previous research,
- the description of the problem should be balanced between generality and specificity,
- the basic concepts, constructs and relations of the theory should be defined clearly and precisely,
- the difference between previously known theoretical work and new theory should be significant.

Uttal (2012), based on the works of Rose (1954), Popper (1959), and Kuhn (1996), proposed the following list of basic requirements that a scientific theory should meet: it should have accuracy, be broad in scope, possess consistency, cumulativeness, and fruitfulness, have a scientific nature, simplicity, and testability. A slightly different list of requirements for a properly constructed theory is presented in the work by Quine and Ullian (1980): uniqueness, parsimony, conservatism, generalizability, fecundity, internal consistency, empirical riskiness, and abstraction.

In view of such a wide variety of opinions on the formalization of the concept of theory, we propose to define this concept based on the requirements that a properly constructed scientific theory should meet. Based on the opinions quoted above, we have distinguished 11 of the most important characteristics of a properly constructed theory and assigned them to five basic groups: structure, scope, capacity to change, empirical compliance, and cognitive values. A diagram of this concept is shown in Figure 11.1.

The *structure* of the theory should ensure its scientific nature by using both scientific language of description and applying scientific methods. Thus, the description should be *abstract*, i.e., objective, and universal, the structure of the theory should ensure *internal coherence* and integrity, and the whole theory should organize knowledge in each area by explaining

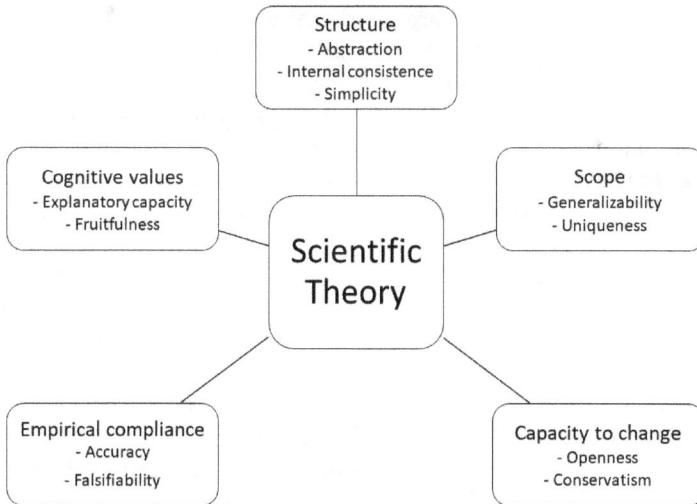

**Figure 11.1:** Key features of scientific theories.

as many observations as possible with the fewest possible assumptions and statements (*simplicity*).

Broad in scope: A theory must have something to say about a domain of sufficient breadth and thus extend beyond particulars to generalizations.

Any scientific theory should be as broad in *scope* as possible, i.e., it should be able to *generalize* events or observations. On the other hand, however, it should be *unique* and therefore clearly distinguishable from other accepted theories. The ability of a theory to *change* is determined on the one hand by its cumulative nature, i.e., its *openness* to new facts (data, information, and phenomena), but on the other hand by its *conservative* nature, i.e., its resistance to possible disturbances. This means that the current version of a theory should not be replaced by a new version if the superiority of this new theory is not clearly demonstrated.

The *empirical compliance* of a theory requires that all theory's predictions should agree with empirical observations (*accuracy*). On the other hand, a properly constructed theory must be susceptible to the risk of being *falsified* by appropriate tests, i.e., a theory must be open to empirical examination. Therefore, it must be refutable. The usefulness of a theory is predominantly determined by its *cognitive values*, sometimes also called epistemic or constitutive values. These include above all the ability to explain reality (*explanatory capacity*), i.e., to answer questions such as who, what, when, where, how and why, as well as *fruitfulness*, i.e., the ability to support prediction and understanding other observations and extend the limits of our knowledge, e.g., by generating new hypotheses and models. However, for the purposes of further considerations it is assumed the simplest of the definitions, proposed by Popper in 2005, namely: "Scientific theories are universal statements. Like all linguistic representations they are systems of signs or symbols."

From a methodological point of view, a distinction is generally made between formal and real sciences. The basis for this division is the nature of the language used in each science, specifically its form and content. The language of formal sciences (which includes mathematics and logic) has no real content and consists of tautologies. The language of the real sciences, on the other hand, is constructed of empirical sentences, and therefore capable of content interpretation in the real world. The real sciences are divided into the empirical and the humanities. It follows that the area of knowledge concerning cognitive dependability could be in *empirical real sciences*.

Knowledge is understood as the totality of convictions from the set of human opinions that are true in the light of facts and reason, whereby irrational knowledge (e.g., mysticism) and rational knowledge, which should be verifiable and communicable, are distinguished. *Rational knowledge* can be divided into:

- colloquial, which is general, ambiguous, and imprecise,
- artistic and literary, which includes literature and art,
- speculative, which includes all speculative systems, such as philosophy or religion,
- scientific, which meets the so-called *strong principle of rationality* (the degree of conviction of the proclaimed belief should correspond to the degree of its justification).

The basic conditions that scientific knowledge must fulfil are logical conformity, both formal and methodological, as well as sociological conformity, generally interpreted as the acceptance of most of the scientific community in each area of knowledge. The conditions of scientific cognition can be reduced to the following basic formal requirements:

- generality as to the scope of the subject of the statement,
- accuracy of the adjudicator of the statement,
- information content (the greater the generality and accuracy, the greater the information content),
- epistemological certainty, understood as the degree of empirical confirmation,
- logical simplicity, measured by the degree of logical systematization of the tested relations.

Relationships between individual requirements are represented by relational models. Figure 11.2 presents a diagram of these relationships based on the Einstein's approach (https://plato.stanford.edu/entries/einstein-philscience/).

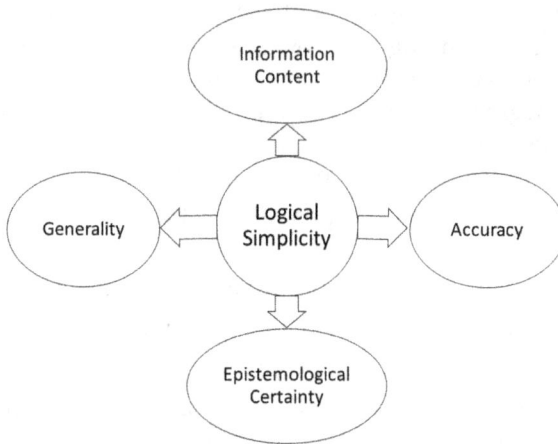

**Figure 11.2:** Conditions of scientific cognition according to the Einstein's model.

Based on widely accepted theories of the evolution of science (Lakatos and Musgrave, 1975; Kuhn, 1996; Popper, 2005), *the process of emergence of a new science* can be divided into three main stages and generally described as follows:

Step I – Conceptual level

1. Every science begins its existence by specifying the basic concepts it will use.

2. Distinguishing a new science requires defining basic features that distinguish it from other, already existing sciences, at least in terms of:

   a. *The subject of research*, i.e., the scope of cognitive interests,

   b. *Research perspective*, i.e., the point of view from which the research subject will be observed and analysed,

   c. *Research methods*, whereby the first stage of the new science is characterized by a high degree of "borrowing" of methods from sciences already fully formed.

3. The basic characteristics of the new science should make it possible to define the boundaries between the new science and the already existing sciences, but these boundaries must not be sharp because:

   a. Research areas of different sciences overlap,

   b. The development of sciences causes information flows between disciplines and fields of science, and even areas of knowledge.

Step II – Level of formalization

4. Each science should strive to create theorems, laws, and scientific theories, looking for regularities in the structure and behaviour of the studied part of reality.

5. According to the concept of development of science by Thomas S. Kuhn, the beginning of a new science can be said to occur when at least one essential *paradigm* is formulated. Paradigms can have different degrees of generality, depending on whether they concern entire fields of science (the so-called macro-paradigms) or a scientific discipline (the so-called micro-paradigms).

6. The formulation of paradigms should be based on the following principles:

   a. Consistency with reality (conformity with available observations),

   b. Rational generality (as broad a scope as possible),

   c. Relevance (cognitive validity for the new science).

7. At the initial stage of the development of a new scientific discipline, paradigms can be temporary, i.e., play the role of so-called "pre-paradigms".

Step III – Application level

8. Every science should fulfil the following cognitive functions:

   a. Descriptive,

   b. Explanatory,

   c. Predictive (prognostic) and in applied sciences,

   d. Prescriptive.

Thus, Cognitive Dependability Engineering (CDE) as a field of scientific knowledge could be considered an empirical applied science and meets all the above-mentioned conditions of scientific cognition.

According to the model of a new science emergence process, the next Section (11.2) attempts to define the theoretical foundations of CDE at a conceptual level, while the next Section (11.3) will focus on formalizing the concept in the form of a paradigm and a framework based on it. Step III, the application level of the CDE concept, will be developed in Part IV in Chapter 17.

## 11.2 Specifying Fundamentals of Cognitive Dependability Engineering

The general concept of Cognitive Dependability Engineering (CDE) is based on the following assumptions:

a. The subjects of CDE research are complex systems in which humans with their cognitive abilities play a central role. Thus, they are *human-centric cyber-physical-social systems* (H-C C-P-S Systems). Due to the high degree of complexity and the significant level of uncertainty, conventional models used successfully in simpler systems cannot be used to describe the behaviour of these systems. This is especially true for attributes related to the reliability, safety, security, and resilience of these systems. The specifics of these types of systems are discussed in Chapter 5 under the Cognitive Systems Engineering concept.

b. The research perspective from which objects of interest (H-C C-P-S Systems) are observed and analysed is the continuity of operation and functioning. In the case of classical systems engineering, this continuity is described by the term reliability, and in the case of cyber-physical systems by the term dependability. However, the presence in H-C C-P-S type systems of a human agent characterized by cognitive abilities necessitates a broader approach to this problem. Such an approach is characterized, among other things, by considering various aspects of the use of systems, not only operating and functioning, but also meeting the cognitive needs of humans and communities, which are subjective in nature. Therefore, it is proposed to introduce the concept of *trustworthy performance* as the main research aspect representing the research perspective of the H-C C-P-S type systems. The concept of trustworthy performance is the core of the Cognitive Dependability Engineering idea, so it will be discussed in detail in Section 11.3.

c. The research method recommended for such defined cognitive goals and such complex diverse systems is a cross-disciplinary methodology called *transdisciplinary*. The specifics of this methodology will be further characterized below.

The following approaches are distinguished in general research methodology (Jantsch, 1970):

- monodisciplinary – characterized by specialization in one discipline in isolation from others,
- multidisciplinary – with one-level multi-goal, but without cooperation,
- pluri-disciplinary – with one-level multi-goal, cooperation, but no coordination,
- cross-disciplinary – one-level one-goal, and rigid polarization towards a specific monodisciplinary concept,
- interdisciplinary – two-level multi-goal, coordination from higher level,
- transdisciplinary – multilevel multi-goal, coordination of the whole system towards a common goal.

*Transdisciplinary research* (TDR) is a process of mutual learning across disciplines and societal actors aimed at creating new knowledge that benefits both scientific praxis and discourse, as well as societal problems

(Scholz and Steiner, 2015a). There are various conceptualizations of transdisciplinary research, which describe this type of research as a process whereby science and society interact to develop and integrate new knowledge (Fam et al., 2018). The transdisciplinary approach can be divided into three main phases: problem framing; creation of solution-oriented knowledge; and integration of knowledge with scientific, empirical, and societal practice (Scholz and Steiner, 2015b). Each of these phases has several challenges, issues, and obstacles.

The transdisciplinary research process involves 'actors' from different domains to produce action-oriented new knowledge, which has the potential to contribute to desired change and development (Caniglia et al., 2020). The typical transdisciplinary research process highlights that such exploration is constituted in combining societal and scientific practice in a mutual learning process which use innovative formats and methods that go beyond the established disciplinary and interdisciplinary scientific repertoires to foster knowledge integration and cognitive processes. These research aspects are particularly important for knowledge generation procedures in TDR as they all include the complex task to integrate various epistemologies from heterogeneous actors. Context dependencies are defined by the research object and its local embeddedness (e.g., space, time). In some cases, it is important to understand the cultural context of a explorative project to recognize the potential, but also the limits of transferring methods and insights to other research projects and contexts (Nagy et al., 2020). Innovative formats and methods in TDR are different due to the variety of societal problems and usage in various disciplines and therefore different approaches and characteristics developed (Wanner et al., 2018). They are used in many contexts and include different methods, e.g., arts-based ones (Peukert et al., 2021). Societal effects are a fundamental aim of transdisciplinarity. However, the variety of terms, concepts, approaches, tools, and methods as well as the difficulty to attribute effects to certain research activities are still challenging for capturing and fostering societal effects. Scientific effects are the fundamental aim of TDR, which are usually difficult to measure. These must be balanced with societal effects which are often more important (Newig et al., 2019). While these aspects, except scientific effects, are often addressed separately, there is still a lack of systematization, operationalization, and understanding how to address and prepare researchers for the interconnections between these four key aspects adequately (Lam et al., 2021).

The main goal of transdisciplinary research is to acquire new integrated knowledge. A simplified diagram of this process is shown in Figure 11.3, which uses the concept of scientific theory (see Figure 11.1) and the trends in TDR discussed above.

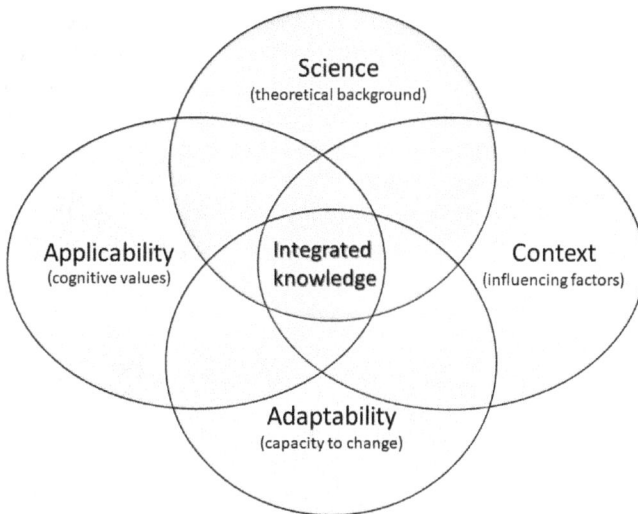

**Figure 11.3:** Diagram of the transdisciplinary research process.

## 11.3 Formalization of the Concept of Cognitive Dependability Engineering

For such a complex research object as Cognitive Systems Engineering, the problem of concept formalization is extremely difficult and requires the consideration of many interdependent factors. This problem is described in a publication by R. Kumar (2011), as following:

**Theoretical framework: As you start reading the literature, you will soon discover that the problem you wish to investigate has its roots in a number of theories that have been developed from different perspectives. The information obtained from different sources needs to be sorted under the main themes and theories, highlighting agreements and disagreements among the authors. This process of structuring a 'network' of these theories that directly or indirectly has a bearing on your research topic is called the theoretical framework.**

As shown in Chapter 10, the quantitative methods used within Dependability Engineering are based on the tacit assumption that the data needed to estimate dependability indices are available in the quantity necessary to produce results of acceptable accuracy. In practice, this assumption is often not met, so a reasonable approach is to consider the effect of the size of the set of available reliability data on the strength of knowledge about the system under study, the level of uncertainty, and the resulting dependability evaluation method. Figure 11.4 shows such relationships by distinguishing four areas of data set size, and the limiting numbers for these intervals are: 1, 30 (the lower value of the so-called representative sample in statistics), and 1000. It can be seen from the figure that uncertainty has a direct impact on how dependability is evaluated, and therefore is an important factor that should be considered when selecting an evaluation method. The second important factor is the complexity of the system being evaluated. In general, the higher the degree of complexity, the more difficult it is to assess the reliability of the system using classical estimation methods. The influence of these two factors on the way dependability aspects are viewed in engineered systems is shown in Figure 11.5.

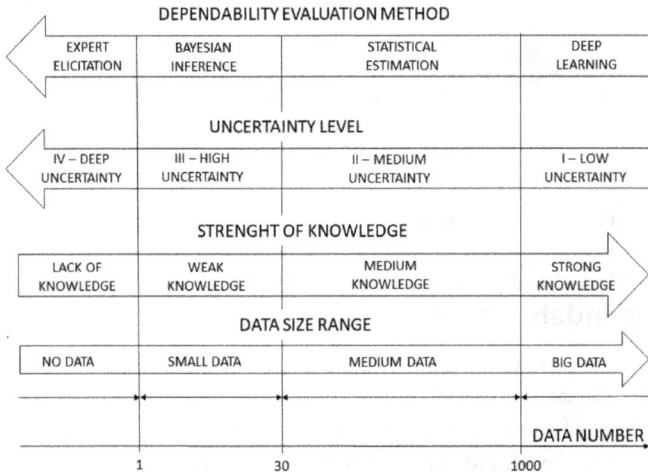

**Figure 11.4:** Relationships between data number, data size range, strength of knowledge, uncertainty level, and dependability evaluation method.

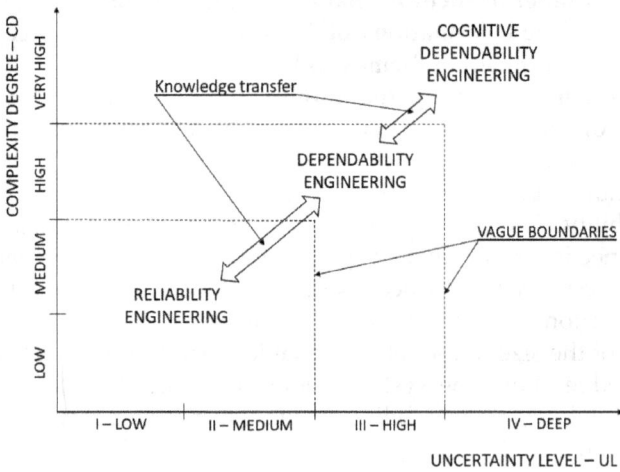

**Figure 11.5:** Areas of applicability of specific reliability/dependability engineering depending on the level of uncertainty and complexity of the system.

The complexity degree (CD) scale can be interpreted as follows:

- Low – simple physical systems,
- Medium – sophisticated physical systems,
- High – cyber-physical systems,
- Very high – cyber-physical-social systems.

Figure 11.5 shows that if the uncertainty is deep and/or the complexity is very high, the recommended method for assessing the dependability of systems should be in Cognitive Dependability Engineering. This is due to two considerations:

- in the case of deep uncertainty, the only possible evaluation method is expert elicitation, which is highly dependent on the cognitive abilities of the experts;
- the structures of systems of very high complexity are so complicated that known models cannot be used to describe them, but only intuitive models based on subjective expert experience.

Thus, the concept of Cognitive Dependability Engineering is based on cognitive processes leading to subjective, largely intuitive expert judgement, which in engineering practice can be equated with elicitation processes. Therefore, the next part of the section will be devoted to the specifics of elicitation processes.

In general, the concept of probability can be understood as an expression of uncertainty about an unknown quantity or state of nature (Berger, 2004). There are two main ways of expressing probability: frequentist and subjective. Frequency concepts assume the recurrence of phenomena and events under stable conditions and boil down to an assessment of the number of occurrences of a given outcome over a specified time interval. Thus, it is an idealization of real conditions, allowing the repetition of a research experiment in an arbitrarily longtime interval. Proponents of the frequency approach interpret probability as a property of the external world, i.e., limiting the relative frequency of occurrence of an event when the number of appropriately defined trials approaches infinity (David and Galavotti, 2009). The assumptions on which the frequency approach is based can therefore be reduced to the following three postulates (Wasserman, 2006):

- Probability refers to an assessment of the relative frequencies of occurrence of an event. Probabilities are objective properties of the real world.
- The parameters of probability distributions are unknown constants. Because they do not fluctuate, useful probability statements cannot be made about the parameters themselves.

- Statistical procedures should be designed to have well-defined frequency properties over the longest possible time horizon.

For the subjectivist, probability is interpreted as the degree of an individual's (e.g., an observer of a certain phenomenon or event) internal belief, as opposed to objective features of the external world. Subjective probability measures the relationship between an observer and events (not necessarily repeatable) in the external world, expressing the observer's personal uncertainty about those events. Thus, a subjective probability estimate is an expression of our state of knowledge at a given time about a given topic. Estimating probabilities requires inductive reasoning (weighing evidence and arguments, evaluating uncertainty, and estimating a number based on available information).

Therefore, the belief-type (subjective) interpretation differs from the frequency-based viewpoint in two main respects. First, the belief-based interpretation of probability does not require that the target event be repeatable. Second, probability is not viewed as an objective property of the real World (the observed system), but as a property of the individual person who perceives the real World in a certain way. Thus, differences in the probabilities assigned to a phenomenon or event reflect differences in the personal knowledge of the evaluating individuals. This approach is consistent with the concept of imperfect knowledge, which in practice is always to some extent incomplete and uncertain (Bukowski, 2019). The subjectivist approach to uncertainty assessment is based on Bayes models and is called Bayesian inference, and is based on the following assumptions (Wasserman, 2006):

- Subjective probability describes the degree of the evaluator's belief about a phenomenon or event. Therefore, one can evaluate the probability of any event, not just those that are subject to random variation. It does not refer to any cut-off frequency – it reflects the strength of the belief that a given claim is true.
- One can make probability theorems about parameters, even if they are constants.
- One can make inferences about a parameter A by creating a probability distribution for A. Inferences, such as point estimates and interval estimates, can then be drawn from this distribution.

In general, Bayesian methods are not guaranteed to accurately assess uncertainty, but in many cases, especially in Big Data mining and machine learning, they are currently essential.

The subjective probability should reflect the initial state of knowledge about some object of interest (prior probability distribution), from which a rational person, using Bayesian methodology, would form an updated posterior probability distribution with the new data available. Bayesian reasoning thus requires two steps: first, deriving the prior distribution, and then modifying that distribution based on the new data obtained. As interest in this issue has grown, more and more work has appeared in the field of psychology on human decision-making under uncertainty. Experiments typically involved asking questions to which subjects did not know the answer, and then asking them to quantify their uncertainty about their answers. Typically, this uncertainty was expressed in terms of probabilities by the psychologists who compiled the findings. This research found that human beings are subject to several errors that distort their judgement about the uncertainty of their knowledge. The commonly cited errors were as follows (Galway, 2007):

- accessibility – the tendency to overestimate the likelihood of events that were easy to recall,
- representativeness – judging the probability of events by focusing on irrelevant features that happen to resemble other events known from the past,
- anchoring – the tendency to maintain an initial assessment of probability value by ignoring new evidence that often contradicts that assessment,
- overconfidence – underestimating uncertainty about a given quantity because of overestimating one's own knowledge about it.

To minimize the impact of these human limitations on the results of their assessments, a new approach to subjective estimation of uncertainties called elicitation has been proposed. The following definitions are proposed:

Elicitation – collection of information based on interviews with people to acquire or supplement knowledge on a specific topic.

Expert elicitation – synthesis of subjective opinions of experts obtained by interview method on a specific topic, used when there is uncertainty about it (topic) due to insufficient objective data.

The choice of the elicitation technique depends on the available resources and the type of information that must be extracted. It can be distinguished between several classes of elicitation techniques (based on https://www.sciencedirect.com/topics/computer-science/elicitation-process):

- Traditional techniques – include a broad class of general data collection techniques, such as the use of questionnaires and surveys, interviews, and analysis of existing documentation.
- Group elicitation techniques – are designed to foster consensus and stakeholder engagement, while using team dynamics to gain a richer understanding of needs. These include brainstorming and focus

groups, as well as consensus-building workshops with an impartial facilitator.

- Prototyping – used for elicitation when there is a high level of uncertainty in the topic area under study or when early stakeholder feedback is needed. Prototyping can also be combined with other techniques, e.g., by using a template (pattern) to provoke discussion during a group elicitation meeting or as the basis for a 'think-aloud' questionnaire or protocol.

- Model-driven techniques – provide a specific information model that determines the entire elicitation process, such as goal-based and scenario-based methods.

- Cognitive techniques – include a range of techniques developed to extract knowledge for knowledge-based systems, such as cognitive protocol analysis (an expert thinks aloud while performing a task to provide an observer with insight into the cognitive processes used to perform the task), laddering (using elicitation to identify the structure and content of stakeholders' knowledge), card sorting (asking stakeholders to sort cards according to their membership in relevant groups), and repertoire grids (constructing attribute matrices for individual objects).

- Contextual techniques – an alternative to both traditional and cognitive techniques. These include participant observation and conversational analysis, used to identify recurrent patterns, such as in conversation and interaction.

Summarizing the literature on practical principles for applying elicitation techniques (e.g., Morgan and Henrion, 1990; Galway, 2007), it can be stated that:

- Experts selected for elicitation should have both the technical knowledge to understand complex technological issues and sufficient management experience to appreciate the organizational and economic challenges that may arise in the future and be independent of the principal.

- The elicitation process is often rushed due to expert time constraints and financial limits. In some cases, the evaluation process is conducted remotely (e.g., using a web-based form), with little interaction between the experts and the object being evaluated.

- The methods used in elicitation processes are mostly ad hoc and rarely based on scientific grounds. In addition, the experts are not informed about the elicitation results or even about the historical data.

- There is mostly a lack of reliable documentation about the elicitation process itself, as well as the forms used. It is especially difficult to go back to completed projects and obtain historical information on past studies. As a result, it is almost impossible to do a comparative analysis of how accurate the various survey methods were in capturing the results.

The rapid development of risk management methods requires quantifying the probability of hazardous events, which may be extremely rare or the result of novel technologies. This has spurred the development of new expert elicitation methods, e.g., an extensive study of nuclear reactor safety analysed the effectiveness of the expert elicitation methods used to assess the probability of types of hazards (Wheeler et al., 1989). This type of methodology was later extended to the area of environmental risk, which required quantifying the potential for environmental hazards to occur, the exposure of various populations to these hazards, and the health effects of these exposures on individuals (Morgan and Henrion, 1990).

An interesting example of the practical application of experience from multiple areas of elicitation application is the paper entitled 'Subjective Probability. Best Practices in Dam and Levee Safety Risk Analysis, 2019'(https://www.usbr.gov/ssle/damsafety/risk/BestPractices/Presentations/A6-subjectiveProbabilityPP.pdf) describing best practices in dam and levee safety risk analysis, which provides questions that experts should ask when assessing the probability of safety risks to critical infrastructure systems. These include the following questions:

- What is the lowest probable value you can think of?
- What is the highest probable value you can think of?
- Is it more likely that the actual value is between these values?
- If so, what is the most likely value?

The current best practice of eliciting a subjective probability distribution from expert opinion can be reduced to the following procedure, which is a synthesis of the experience of many experts in the field (e.g., Morgan and Henrion, 1990; Chaloner, 1996; Meyer and Booker, 2001; Galway, 2007):

- Using the knowledge of multiple independent experts, following the principle, the more the better.
- Not limiting oneself to estimating a single value (e.g., most likely), but asking experts to provide three values (e.g., upper, lower, and most likely) for the parameter under consideration and justifying the values provided.
- It is suggested that the central (or most likely) value be evaluated last to counteract the anchoring effect.

- Provide feedback to the experts on the final evaluation results, preferably at the end of the elicitation session. Such feedback would include all data on the evaluation results by all experts and provide an opportunity to discuss the most controversial evaluations.

- Carefully documenting the entire elicitation process, describing the results obtained, and archiving the data for future retrospective studies.

A practical example of the use of elicitation is presented in the paper "A method for estimating the probability distribution of the lifetime for new technical equipment based on expert judgement" (Andrzejczak and Bukowski, 2021).

Based on the above considerations, the following definitions are proposed:

COGNITIVE DEPENDABILITY (CD) IS THE ABILITY OF HUMAN-CENTRIC CYBER-PHYSICAL-SOCIAL SYSTEMS (H-C C-P-S SYSTEMS) TO PERFORM TRUSTWORTHY IN A VARIETY OF SITUATIONS, INCLUDING CONDITIONS OF DEEP UNCERTAINTY.

COGNITIVE DEPENDABILITY ENGINEERING (CDE) IS A GENERALIZED VERSION OF THE DEPENDABILITY ENGINEERING TARGETING COMPLEX HUMAN-CENTRIC CYBER-PHYSICAL-SOCIAL SYSTEMS WHICH SHOULD PERFORM TRUSTWORTHY IN A VARIETY OF SITUATIONS, INCLUDING CONDITIONS OF DEEP UNCERTAINTY.

Accepting the above definitions, the following *pre-paradigm* is suggested as a starting point for further research in CDE:

THE CONCEPT OF COGNITIVE DEPENDABILITY ENGINEERING IS BASED ON COGNITIVE PROCESSES LEADING TO SUBJECTIVE, LARGELY INSTINCTIVE EXPERT JUDGMENT, WHICH IN ENGINEERING PRACTICE MAY BE SUPPORTED BY INTUITIVE MODELS AND ELICITATION PROCEDURES.

In compliance with the above definition and pre-paradigm, Chapter 16 presents a framework for practical use of the Cognitive Dependability concept for managerial decision support.

## 11.4 References

Andrzejczak, K. and Bukowski, L. (2021). A method for estimating the probability distribution of the lifetime for new technical equipment based on expert judgement. *Eksploatacja i Niezawodność – Maintenance and Reliability*, 23(4): 757–769. http://doi.org/10.17531/ein.2021.4.18.

Berger, J. (2004). The case for objective Bayesian Analysis. *Bayesian Analysis*, 1(1): 1–17.

Bukowski, L. (2019). *Reliable, Secure and Resilient Logistics Networks. Delivering Products in a Risky Environment*, Springer Nature Switzerland AG 2019, ISBN 978-3-030-00849-9 (Hardcover), ISBN 978-3-030-00850-5 (eBook).

Caniglia, G. et al. (2020). A pluralistic and integrated approach to action-oriented knowledge for sustainability. *Nature Sustainability*, 4: 93–100. https://doi.org/10.1038/s41893-020-00616-z.

Chaloner, K. (1996). Elicitation of prior distributions. *In*: Berry, D.A. and Stangl, D.K. (eds.). *Bayesian Biostatistics*, New York: Marcel Dekker.

Dawid, A.P. and Galavotti, M.C. (2009). De Finetti's subjectivism, objective probability, and the empirical validation of probability assessments. pp. 97–114. *In*: Galavotti, M.C. (ed.). *Bruno de Finetti Radical Probabilist*. London: College Publications.

Einstein's Philosophy of Science. https://plato.stanford.edu/entries/einstein-philscience/.

Fam, D., Neuhauser, L. and Gibbs, P. (eds.). (2018). *Transdisciplinary Theory, Practice and Education. The Art of Collaborative 600 Research and Collective Learning*. Cham, Switzerland: Springer. https://doi.org/10.1007/978-3-319-93743-4.

Feldman, D.C. (2004). What are we talking about when we talk about theory? *Journal of Management*, 30(5): 565–567.

Galway, L.A. (2007). Subjective Probability Distribution Elicitation in Cost Risk Analysis, RAND Corporation. https://www.rand.org/content/dam/rand/pubs/technical_reports/2007/RAND_TR410.pdf.

Hooker, C.A. (1987). *The Realistic Theory of Science*. State University of New York. https://www.sciencedirect.com/topics/computer-science/elicitation-process.

Hull, C.L. (1943). *Principles of Behavior: An Introduction to Behavior Theory*. Appleton-Century.

Jantsch, E. (1970). Inter-disciplinary and transdisciplinary university. Systems approach to education and innovation. *Policy Sci.*, 1: 403–428.

Kaplan, A. (1964). *The Conduct of Inquiry: Methodology for Behavioral Science*. S.F. USA: Chandler Publishing Company.

Kerlinger, F.N. (1986). *Foundations of Behavioral Research*. New York, Holt: Rinehart and Winston.

Kuhn, T.S. (1996). *The Structure of Scientific Revolutions*. The University of Chicago Press.

Kumar, R. (2011). *Research Methodology: A Step-by-step Guide for Beginners*. 3rd Edition. New Delhi: Sage.

Lakatos, I. and Musgrave, A. (eds.). (1970). *Criticism and the Growth of Knowledge*. Cambridge University Press.

Lam, D.P.M. et al. (2021). Transdisciplinary research: Towards an integrative perspective. *GAIA*, 30/4: 243–249.

Merriam-Webster Dictionary, https://www.merriam-webster.com/dictionary/theory.

Meyer, M.A. and Booker, J.M. (2001). *Eliciting and Analyzing Expert Judgment: A Practical Guide*. Philadelphia, Pa.: Society for Industrial and Applied Mathematics and the American Statistical Association.

Morgan, M.G. and Henrion, M. (1990). *Uncertainty: A Guide to Dealing with Uncertainty in Quantitative Risk and Policy Analysis*. New York: Cambridge University Press.

Nagy, E. et al. 2020. Transfer as a reciprocal process: How to foster receptivity to results of transdisciplinary research. *Environmental Science and Policy*, 104: 148–160. https://doi.org/10.1016/j.envsci.2019.11.007.

Newig, J., Jahn, S., Lang, D.J., Kahle, J. and Bergmann, M. (2019). Linking modes of research to their scientific and societal outcomes. Evidence from 81 sustainability-oriented research projects. *Environmental Science and Policy*, 101: 147–155. https://doi.org/10.1016/j.envsci.2019.08.008.

Peukert, D., Lam, D.P.M., Horcea-Milcu, A.I. and Lang, D.J. (2021). Facilitating collaborative processes in transdisciplinary research using design prototyping. *Journal of Design Research*, 18/5/6: 294–326. https://dx.doi.org/10.1504/jdr.2021.10041108.

Popper, K. (1959). *The Logic of Scientific Discovery*. Hutchinson & Co.

Popper, K. (2005). *The Logic of Scientific Discovery*. Taylor & Francis e-Library. http://strangebeautiful.com/other-texts/popper-logic-scientific-discovery.pdf.

Quine, W.V. and Ullian, J.S. (1980). Hypothesis. *In*: Klempe, E.D., Hollinger, R. and Kline, D.A. (eds.). *Introductory Readings in the Philosophy of Science*. Buffalo, New York: Prometheus Books.

Rose, A.M. (1954). *Theory and Method in the Social Sciences*. University of Minnesota Press.

Rudner, R.S. (1966). *Philosophy of Social Science*. Englewood Cliffs, NJ: Prentice-Hall.

Scholz, R.W. and Steiner, G. (2015a). The real type and ideal type of transdisciplinary processes: Part I—Theoretical foundations. *Sustainability Science*, 10(4): 527–544.

Scholz, R.W. and Steiner, G. (2015b). The real type and ideal type of transdisciplinary processes: Part II—What constraints and obstacles do we meet in practice? *Sustainability Science*, 10(4): 653–671.

Subjective Probability. *Best Practices in Dam and Levee Safety Risk Analysis*. (2019). (https://www.usbr.gov/ssle/damsafety/risk/BestPractices/Presentations/A6-subjectiveProbabilityPP.pdf).

Uttal, W.R. (2012). *Neural Theories of Mind: Why the Mind–Brain Problem May Never Be Solved*. New York, London: Psychology Press, Taylor and Francis Group.

Wacker, J.G. (1998). A definition of theory: Research guidelines for different theory-building research methods in operations management. *Journal of Operations Management*, 16: 361–385.

Wanner, M., Hilger, A., Westerkowski, J., Rose, M., Stelzer, F. and Schäpke, N. (2018). Towards a cyclical concept of real-world laboratories. A transdisciplinary research practice for sustainability transitions. *disP – The Planning Review*, 54/2: 94 –114. https://doi.org/10.1080/02513625.2018.1487651.

Wasserman, L. (2006). Frequentist Bayes is objective (Comment on articles by Berger and by goldstein). *Bayesian Analysis*, 1(3): 451–456.

Wheeler, T.A., Hora, S.C., Cramond, W.R. and Unwin, S.D. (1989). *Analysis of Core Damage Frequency from Internal Events: Expert Judgment Elicitation*. Vol. 2, Washington, D.C.: Nuclear Regulatory Commission, NUREG/CR-4550.

Part III

# Modelling and Simulation the Operation of Cyber-Physical-Social Systems in a Risky Environment

# 12

# Methodology of Modelling and Simulation Used for Complex Systems*

Chapter 12 provides the fundamentals of modelling and simulation of complex engineered systems. A model is a surrogate for the real system and represents its characteristics and attributes in experimental studies. Modelling process is a constructive activity whose goal is to build a sufficiently good model. The chapter presents the algorithm of the entire modelling and simulation process and discusses all its steps. Then the process of creating mathematical models divided into seven stages is demonstrated. The basic advantages and disadvantages of the modelling and simulation process were examined. The second part of the chapter was devoted to development and application of the modelling and simulation process. The lifecycle of a modelling and simulation process was presented and analysed. The relationships between the conceptual model, the working model, and the real complex engineered system was discussed. The eight key factors contained in the examination of results credibility were demonstrated and analysed. The chapter concludes with an analysis of the risks associated with uncertainty regarding the credibility of modelling and simulation results, as well as criteria for deciding whether to accept or reject these results.

---
* https://orcid.org/0000-0002-2630-3507.

## 12.1 Modelling and Simulation of Complex Engineered Systems

The phrase 'modelling and simulation' (M&S) has a generally accepted meaning and implies two distinct activities. The modelling activity creates a model which is subsequently used as an object for experimentation. This experimentation with the model is the simulation activity. The term 'model' refers to an abstract representation of the reality. The use of models (especially, mathematical ones) as a basis for analysis and reasoning is well established in many scientific and practical disciplines. The widespread availability of computing power makes the experimentation with complex models more and more promising and therefore, the emergence of the modelling and simulation as a new research field. There is a strong connection between the model that is appropriate for the investigation and the nature of the problem that should be solved. However, there rarely exists the universal model which will support all modelling and simulation projects that have a common system context. This is particularly true when the system has a high level of complexity. Identification of the most appropriate model for complex engineered systems is usually the most challenging aspect of the modelling and simulation approach to solving.

Each model is a surrogate for the real system and represents its characteristics and attributes in experimental studies. When the underlying system does not yet exist, which means it is in the development stage as a certain idea or concept, then modelling is the only option for experimentation. But even when the real system exists there is a variety of reasons why experimenting directly with it could be inappropriate. For example, such experimentation can be limited by (based on Birta and Arbez, 2007):

- Costs – e.g., determining the performance and reliability by upgrading the system at all the nodes of a large communications network;
- Safety considerations – e.g., if operational events pose a serious safety risk an experiment could be too dangerous;
- Time constraints – e.g., investigation of slowly occurring changes could be too time-consuming;
- Possible consequences – e.g., too disruptive because of cascading failures;
- Ethical rules – e.g., experiments on people and their feelings are ethically unacceptable;
- Irreversibility of consequences – e.g., investigating the impact of a fiscal policy change on the economy of a country.

A typical model includes specifications for system behaviour and the modelling process is focused on the development of this specification. The goal of this process is to ensure that the behaviour of the model will be as similar as possible to the behaviour of the real system. The main challenge is to capture all relevant details and to avoid superfluous features, which is in line with the well-known quotation from Albert Einstein: "Everything should be made as simple as possible, but not simpler."

The modelling process is a constructive activity whose goal is to build a sufficiently good model. Typically, a key question is whether the model is good enough from the point of view of the project's objectives. It means that it is not meaningful to undertake any modelling study without a clear understanding of the purpose for which the model will be used. There is a diversity of ways in which the specification of behaviour can be formulated, for example: natural language, mathematical formulas, rule-based formalisms, symbolic or graphical descriptions, and combinations of these. A particular factor that plays an important role is a specification formulated as a computer program, because computer programs provide the means for carrying out the experiments that are central to the modelling and simulation approach.

Nevertheless, like other methods, modelling and simulation have some weaknesses, and must be used with great care and caution. Here are typical errors that can lead to failure in the modelling and simulation process (based on Birta and Arbez, 2007):

- Incorrect statement of goals. No project can ever be successful unless its objectives are clearly articulated and fully understood by all the stakeholders. Ambiguity in the statement of goals can lead to wasted effort or yield conclusions that are unrelated to the objective of the project. The project's goals must be consistent with the maturity of knowledge that characterizes the modelled system, as well as the available level of resources (e.g., time, skills, equipment, etc.) should be adequate to achieve the goals.

- Inappropriate granularity of the model. The granularity of the model refers to the level of detail (the degree of resolution) with which it attempts to replicate the system. The level of granularity is necessarily bounded by the goals of the project. Excessive detail increases complexity and this can lead to cost overruns and/or completion delays that usually translate into a project's failure. However, too little detail can mask the very effects that have substantial relevance to the behaviour of the modelled system.

- Ignoring unexpected results. Although a validation process is recognized to be an essential stage in any modelling and simulation project, its main thrust is to confirm that expected behaviour does occur. Such a result of simulation can sometimes occur and when it is observed there often is a tendency to dismiss it, particularly when validation tests have provided satisfactory results. Ignoring such counterintuitive observations can be the cause of failure.

- Unsuitable combination of essential skills. A modelling and simulation project has substantial requirements in terms of both the range of skills and the effort needed for its completion. Team members contribute complementary expertise to the intrinsically multifaceted requirements of the project. The range of skills that needs to be represented among the experts can include project management, documentation, using domain knowledge for building dynamic model, experiment design, software development, and analysis of results. The intensity of coverage of these various areas is dependent on the specific nature of the project. However, an inappropriate mix of skills can ultimately result in project failure.

- Inadequate flow of information. It is necessary to ensure stable and correct flow of information between all team members who implement the project. For example, a minor misinterpretation of requirements can have consequences that can lead to the failure of the entire project.

The typical simulation model consists of the following components: system entities, input variables, performance measures, and functional relationships. Practically all simulation software packages provide constructs to model each of the above components. Modelling is regarded as the most important part of a simulation study, because a simulation study can be only as good as the simulation model. Full simulation process comprises the following 12 steps (based on Maria, 1997 and Bukowski, 2019) shown in Figure 12.1:

Step 1. Problem identification. Enumerate problems with an existing system and produce requirements for a proposed system.

Step 2. Problem formulation. Select the boundaries of the system to be studied and define overall objective of the study as well as some specific issues to be addressed. Define quantitative criteria based on which different system configurations will be compared and ranked. Briefly identify the configurations of interest and formulate hypotheses about system performance. Decide the time frame of the study and identify the end user of the simulation model. All these problems must be formulated as unambiguously and clearly as possible.

Step 3. Data collection and processing. Gather data on system specifications, input variables, as well as performance of the existing system. Identify sources of uncertainty in the system. Select an appropriate input probability distribution for each stochastic input variable and estimate corresponding parameters. Empirical distributions are used when standard distributions

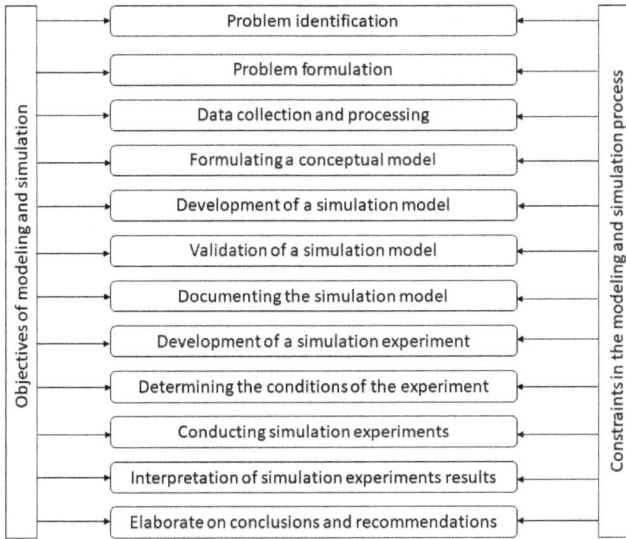

**Figure 12.1:** Diagram of a typical modelling and simulation process.

are not appropriate or do not fit the available system data. Triangular, uniform, or normal distribution is used as a first presumption when no data are available.

Step 4. Formulating a conceptual model. Develop graphics and network diagrams of the system. Translate these *conceptual models* to an acceptable form for the selected simulation software.

Step 5. Development of a simulation model. Verify that the simulation model executes as intended. Verification techniques include traces, varying input parameters over their acceptable range and checking the output, substituting constants for variables, and checking the plausibility of the results.

Step 6. Validation of a simulation model. Compare the model's performance under known conditions with the performance of the real system under the same conditions. Perform statistical inference tests and get the model examined by system experts. Assess the confidence that the end-user places on the model and address eventual problems.

Step 7. Documenting the simulation model. Record objectives and assumptions of the project as well as input variables and the model's parameters in detail.

Step 8. Development of a simulation experiment. Choose a performance measure, a few numbers of input variables that are likely to influence it, and the levels of each input variable. When the number of possible configurations (product of the number of input variables and the levels of each input variable) is large and the simulation model is very complex, common second-order design classes including central composite or full-factorial should be considered. Document all details of the experimental design.

Step 9. Determining the conditions of the experiment. Address the question of obtaining accurate information and the maximum possible results from each run. Determine if the system's behaviour is stationary (performance measure does not change over time) or non-stationary (performance measure changes over time). Generally, in stationary systems, steady-state behaviour of the response variable is of interest. Select the run length, as well as appropriate starting conditions and the length of the warm-up period (so-called transient behaviour), if required. Decide the number of independent runs (each run uses a different random number string and the same starting conditions) by considering output data sample size.

Step 10. Conducting simulation experiments. Runs were performed as specified in Step 9 above. Sample size must be large enough (at least 3–5 runs for each configuration) to provide the required confidence in the performance measure estimates. Check if the output data is not correlated.

Step 11. Interpretation of simulation experiments results. Compute numerical estimates (e.g., mean value, standard deviation, confidence intervals) of the desired performance measure for each configuration of interest. The assumption that batch means are independent may not always be true; increasing total sample size and increasing the batch length may be useful. Test the hypotheses about system performance. Construct graphical displays of the output data. Document results and conclusions.

Step 12. Elaborate on conclusions and recommendations. This may include additional experiments to increase the precision and reduce the bias of estimators (e.g., to perform sensitivity analyses).

The fundamental step in the preparation of the simulation project constitutes the conceptual modelling. *Conceptual modelling* is the process of abstracting a model from a real or proposed new system. The design of the model influences all aspects of the study, in particular the data requirements, the time in which the model can be developed, the validity range of the model, the speed of experimentation, and the confidence and accuracy of experimentation results. Although effective conceptual modelling is a vital aspect of a simulation study, it is probably the most difficult and poorly understood (Robinson, 2008). There are only few publications to the subject of conceptual modelling. The main reason for this situation is because conceptual modelling is more an 'art' than a 'science' and therefore it is difficult to define proper methods and procedures. The result is that the process of conceptual modelling is mostly learnt by experience. The main domain of interest for this chapter is the use of simulation for modelling of complex operating systems. Wild (2002) defined the term 'an operating system' as a configuration of resources combined for the provision of goods or services.' He identifies four specific functions of operations systems: manufacture, transport, supply, and service. Models in these domains tend to be of a relatively small scale, with a project life cycle of normally less than six months (Cochran et al., 1995). Simulation modelling in the military domain as well as for the global logistics networks tend to be of a much larger scale and where they are developed by teams of people.

Usually, a conceptual model is not sufficient to solve a given problem and then there is a need for a more accurate model, which is a mathematical model. The mathematical model is characterized by the highest degree of abstraction and allows the use of universal symbols instead of specific physical quantities. The mathematical model is assumed to be a finite set of symbols and mathematical relations as well as strict operating rules, while the symbols and relations included in the model have an interpretation referring to specific elements of the modelled real system. The set of symbols and relations is an abstract mapping of the system, and the factor transforming it into a mathematical model is its physical interpretation. Generally, *mathematical modelling* is an interdisciplinary field whose task is to describe complex and sophisticated reality in the simplest possible and acceptable way. Thus, the only criterion for the correctness of modelling is the consistency of results obtained with its use with reality identified experimental or with experience.

The process of a mathematical model building is characterized by an iterative procedure, and it can be distinguished in the following seven stages (see Figure 12.2): determining the object of study, establishing research objectives, selecting the model structure, identification of model parameters, development of a computational algorithm, model verification and validation, and acceptance of the model.

Each model should be created for specific systems and phenomena, as well as applications, and therefore the modelling process must be targeted purposeful. Depending on the assumed modelling goal the following types of models are distinguished:

- phenomenological models, whose function is to describe the operation of the system,

- prognostic models, which are used to predict system behaviour in the future under different environmental conditions,

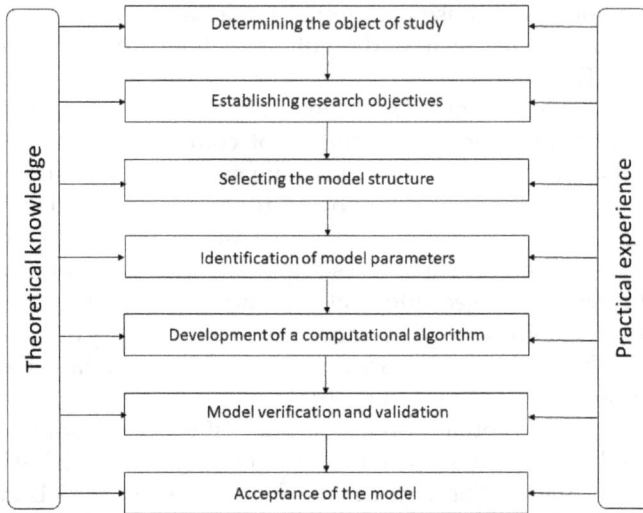

**Figure 12.2:** The process of a mathematical model building.

- decision models, supporting the selection of inputs values, meeting certain conditions at the outputs of the system,
- normative models, supporting the choice of the structure and parameters of the system, fulfilling specific tasks.

The complete system-relevant knowledge about the system should be transformed into a set of consistent mathematical-logical relations. In practice, it is most problematic to find a rational compromise between the degree of complexity of the model and the ease of its use, so-called degree of simplification. Another important problem related to the requirement of model's compliance with the modelled system is the significance problem. It consists in the proper distinction of the essential features of the real system, which must be reflected in the mathematical model. This difficulty usually consists in the lack of a priori appropriate theory and the need to verify it using the results of modelling.

The identification of the model parameters involve the estimation of the numerical values for coefficients occurring in the model. There are two main ways:

- passive identification, consisting in collecting experimental data during the 'normal' operation of the system, and then determining the searched values of the model parameters with statistical methods by estimation, or
- active identification, involving proper planning and conducting an experiment, the results of which will allow to determine the desired parameter values.

Active identification requires more resources and is used to study the behaviour of systems in unusual situations and new operating conditions.

Development of a computational algorithm is directed to choose one of the following approaches: analytical, numerical, or simulation methods. Each of these choices results in a different kind of form of results, namely (Bukowski, 2019):

- Analytical solutions lead to explicit results and allow the assessment of important system properties such as sensitivity and stability.
- Numerical solutions are used when the values of dependent variables can only be determined using appropriate algorithms. The solution algorithm is usually implemented by a computer program.
- Simulation solutions are characterized by the fact that the independent variables of the model correspond to the input quantities, and the dependent variables to the output of the real system. In the case of computer simulation, the numerical procedure is part of the simulation model.

The model verification and validation consist in comparing the results of modelling with the behaviour of the real system due to the compliance with the theoretical knowledge and the results of experimental research. The efficiency of modelling process can be increased by applying verification in all steps of model creation. The compliance of the model with the real system is validated according to the following criteria:

- internal, regarding the features of the model, such as formal compliance (i.e., no logical conflicts) and algorithmic compliance (i.e., the ability to perform calculations with the required accuracy), and
- external, concerning the goals of modelling and compliance of modelled phenomena with theory and experimental data. The heuristic compliance refers to the scientific values of the model, such as: the ability to interpret phenomena, verification of hypotheses, and the pragmatic consistency refers to the results of modelling, verified by comparing these results with the behaviour of the real system under the same conditions.

Acceptance of a verified and validated model also requires a statement of its usefulness, which is the case, for example, in the following cases:

- There is no possibility to observe certain processes in the real world.
- It is not possible or extremely expensive to validate the mathematical model describing the system (e.g., due to insufficient data).
- Although mathematical model can be formulated but analytic solutions are either impossible or too sophisticated (e.g., complex systems like global supply networks).

Most simulation studies are implemented using a simulation software package. The main benefits of this practice are reduced programming requirements, conceptual guidance, automated gathering of statistics, graphic symbolism for user-friendly communication, visual animation, and possibility to change the model. Metrics for evaluation of this packages include modelling flexibility, simplicity (ease for use), a variety of modelling structure (e.g., hierarchical, flat, object-oriented, network and others), code reusability, graphic user interface, hardware and software requirements, statistical capabilities, output reports, customer support, and documentation. The main types of simulation packages are simulation languages and application-oriented simulators. Admittedly, simulation languages offer more flexibility than the application-oriented simulators, but they require varying amounts of programming expertise. Application-oriented simulators are easier to learn and have modelling constructs closely related to the application.

The basic advantages of simulation techniques are (Maria, 1997):

- Employ a system and/or process approach to problem solving.
- Obtain a better understanding of the system by observing the system's operation in detail over any period.
- Test hypotheses about the system for feasibility.
- Compress time to observe slowly changing phenomena over long periods or expand time to observe a fast-changing phenomenon in detail.
- Study the effects of different scenarios on the operation of a system by altering the system's model.
- Experiment with new situations about which only poor information is available.
- Identify the 'driving' variables (ones that performance measures are most sensitive to) and the interrelationships among them.
- Identify bottlenecks in the flow of material, people, or information and weak links in system.
- Use numerous performance metrics for analysing system configurations.
- Develop well designed, reliable systems and reduce system development time.

Nevertheless, simulation can be a time consuming and complex process in the following cases:

- Unclear objective.
- Erroneous assumptions.
- Invalid model.

- Simulation model too complex or too simple.
- Using the wrong input description.
- Replacing a random variable by deterministic value.
- Using the wrong performance measure.
- Not considering an initial bias in output data.
- Making only one simulation run for a configuration.
- Using simulation when an analytic solution is appropriate.

## 12.2 Development and Application of the Modelling and Simulation Process

The typical life cycle of a modelling and simulation (M&S) process has two main steps: M&S development, and model usage. The first step includes model initiation, identification of the model intendent use, concept development, conceptual model design, conceptual validation, model construction, model testing, and model usefulness assessment. The second step of M&S contains model practical application and its archiving. These steps divided into 10 phases are shown in Figure 12.3.

Establishing the need for modelling and simulation starts their life cycle and can occur at any point in a program's/project's development. The need starts the *initiation phase*, where the real complex system (RCS), which is to be modelled, and the information that are required are identified, as well as the intended use of the M&S is defined. The *intended use* is further

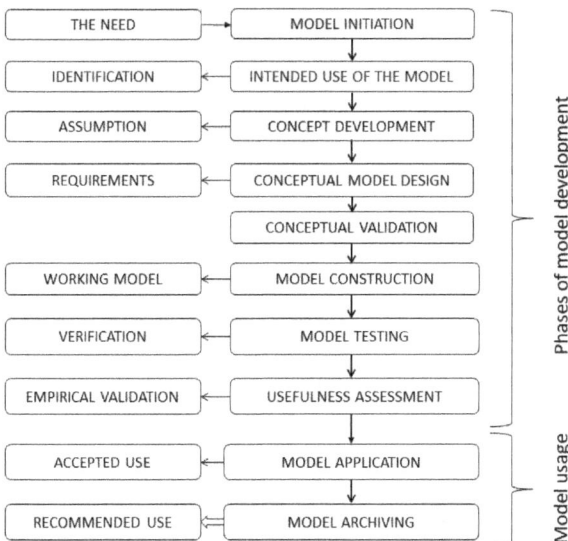

**Figure 12.3:** The life cycle of a modelling and simulation process (based on Bukowski, 2019).

defined during the *concept development phase*, where the aspects of the RCS that should be included in the M&S and the assumptions required to implement the M&S are identified. Next the objectives and level of detail required for the M&S, and the acceptance criteria to determine its sufficiency, are identified. The intended use usually becomes more refined leading to iterations in the M&S development phase in the case of intended use changes.

Throughout the M&S *design phase*, a conceptual model and other requirements or specifications are developed to describe the physical behaviour of the RCS and its interactions with the environment. The conceptual model is validated against the aspects and behaviour of the RCS within the areas of interest as defined by the intended use during the conceptual validation phase. Once the conceptual model is validated, the *working model* is constructed. The relationships between the M&S design (conceptual model), the working model (e.g., computational model), and the RCS as the reality of interest are illustrated in Figure 12.4 (adapted from NASA, 2016).

During the model *testing phase*, verification shows if the working model adequately represents the RCS and behaves similarly to the real system. However, the intent of empirical validation is to identify the model's limits of operation, e.g., the area in which the model behaves correctly. Once at the end of model testing the model is already verified and validated, the model's capabilities, assumptions, and limits of operation are documented and assessed with respect to acceptance criteria to determine the permissible uses of the model. As soon as model testing is

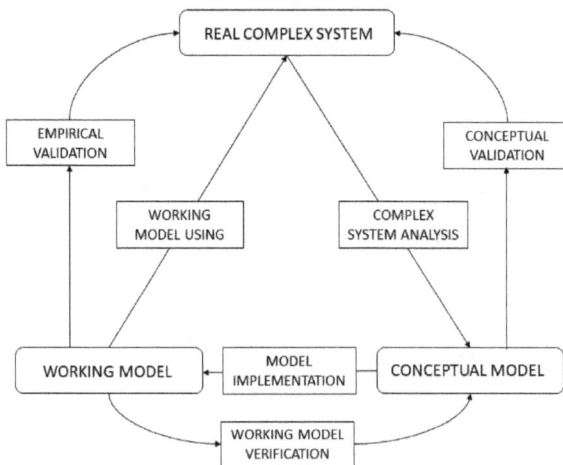

**Figure 12.4:** The relationships between the conceptual model, the working model, and the real complex system (based on NASA, 2016).

successfully completed, the model is released, along with documentation of the model's capability and the range of the model applicability, ending the model development phase.

Throughout the *application phase*, the use of a model starts with an assessment of whether the proposed usage of the model sufficiently matches the permissible ones. If the proposed use is satisfactory, the model is then used to obtain the results of interest. If the proposed model does not meet the defined requirements, the intended use will either be rejected or conditionally allowed with the appropriate restrictions, cautions, or placarding required (NASA, 2016).

An important phase in the life cycle of a modelling and simulation process is the examination of results credibility. The eight key factors contained in this assessment are considered as a minimum set and they are data pedigree, verification, validation, input pedigree, uncertainty characterization, results robustness, history, and process/product management. These factors are grouped into three categories: M&S development (data pedigree, verification, validation), M&S use (input pedigree, uncertainty characterization, results robustness), and Supporting evidence (M&S history, and M&S process/product management), which may span all aspects of an M&S Credibility Assessment Structure. The main aspects assessed by these eight factors include as follows (NASA, 2016):

I. M&S Development

- Data pedigree –to determine whether the pedigree and quality of the data used to develop the model is adequate and acceptable.
- Verification – allows to determine whether the models were implemented correctly, including their requirements and specifications.
- Validation – allows to determine whether the M&S results were close to the referent data and the reality of interest.

II. M&S Operations

- Input pedigree – to determine whether the pedigree and quality of the data used to setup and run the model is adequate or acceptable.
- Uncertainty characterization – to determine whether the uncertainty in the current M&S results is appropriately characterized, as well as establish what are the sources of uncertainty are in the results and how they are propagated through to the results of the analysis.
- Results robustness – to determine how thoroughly are the sensitivities of the current M&S results known.

III. Supporting Evidence

- M&S History – allows to determine how similar is the current version of the M&S to previous versions, and how alike is the current use of the M&S to previous successful uses.
- M&S Management – allows to determine how well managed were the M&S processes and products.

Using a general-purpose software package for modelling and simulation, we must consider that some credibility assessment activities are twofold: those relating to the package itself, and those relating to the implemented M&S process. The former is usually given insufficient attention in the case of M&S using third-party tools, where certain aspects of M&S are neither performed nor managed for some aspects of M&S processes. The constructed computational model using these tools is also evaluated in this category, which necessarily includes the pedigree of the data used in M&S development, as well as verification and validation of the constructed model. However, the M&S Operations category deals with the credibility factors for the application of the specific computational model in the generation of the current simulation results. This includes the present use of the model for the simulation, analysis, and reporting of the results. The use history of both the tool in general and the specific computational model, as well as the overall management of the M&S processes and products involved in the development, operation, and analysis of the computational model, are covered in the Supporting Evidence category (NASA, 2016).

The decision about which cases of modelling and simulation should be the subject to the rules outlined above depends on the result of the M&S risk assessment. Such risk assessments consider both the *consequences* to human safety or project success criteria in case a decision proves incorrect, and the degree to which M&S results *influence* a decision. Decision consequence classifications assess the impact of a decision that proves incorrect and include four different levels (based on NASA, 2013):

- Class IV – negligible. A poor decision may result in the need for minor 'first aid' treatment but would not adversely affect personal safety or health; damage to facilities, equipment, or hardware more than normal wear and tear level; cost overrun less than 2% of planned cost; all criteria met, with at worst minor performance degradation.
- Class III – moderate. A poor decision may result in minor injury, or minor property damage to facilities, systems, equipment, or hardware; cost overrun between 2%–15% of planned cost; moderate performance degradations.

- Class II – critical. A wrong decision may result in severe injury, or major property damage to facilities, systems, equipment, or hardware; cost overrun between 15%–50% of planned; substantial performance degradations.
- Class I – catastrophic. A wrong decision may result in death or permanently disabling injury, facility destruction on the ground, or loss of crew, major systems, or vehicle during the mission; cost overrun greater than 50% of planned cost; most criteria not met due to severe performance degradations.

Influence estimates the degree to which M&S results impact program or project engineering decisions. The decisions include determination of whether design requirements have been verified and include five different levels (based on NASA, 2013):

- Influence 1 – negligible. Results from the M&S are a negligible factor in engineering decisions. This includes research on M&S methods, and M&S used in research projects that have no direct bearing on program/project decisions.
- Influence 2 – minor. M&S results are only a minor factor in any program/project decisions. Test data for the real system in the real environment are available, and M&S results are used just as supplementary information.
- Influence 3 – moderate. M&S results are at most a moderate factor in any program/project decisions. Test data for the real system in the real environment are not available, but ample flight or test data for similar systems in similar environments are available.
- Influence 4 – significant. M&S results are a significant factor in some program/project decisions, but not the sole factor for any program/project decisions. Test data for similar systems in similar environments are available.
- Influence 5 – controlling. M&S results are the controlling factor in some program/project decisions. Test data are available for essential aspects of the system and/or the environment.

The results of the risk assessment can be presented in the form of a risk matrix, an example of which is shown in Figure 12.5. Those M&S that are judged to fall within the red (R) boxes are obligatorily within the scope of the procedure described above, and those that fall within the green (G) boxes are not. The M&S that are referred to fall within the yellow (Y) boxes are recommended to be processed in accordance with the above procedure.

| | IV. NEGLIGIBLE | III. MARGINAL | II. CRITICAL | I. CATASTROPHIC |
|---|---|---|---|---|
| 5. CONTROLLING | G (+) | Y (?) | R (-) | R (-) |
| 4. SIGNIFICANT | G (+) | Y (?) | R (-) | R (-) |
| 3. MODERATE | G (+) | Y (?) | Y (?) | R (-) |
| 2. MINOR | G (+) | G (+) | Y (?) | Y (?) |
| 1. NEGLIGIBLE | G (+) | G (+) | G (+) | G (+) |

RESULTS INFLUENCE

DECISION CONSEQUENCE

**Figure 12.5:** An example of a M&S risk assessment matrix (based on NASA, 2013).

## 12.3 References

Birta, L.G. and Arbez, G. (2007). *Modelling and Simulation Exploring Dynamic System Behaviour.* London: Springer-Verlag London Limited.

Bukowski, L. (2019). *Reliable, Secure and Resilient Logistics Networks. Delivering Products in a Risky Environment.* Springer Nature Switzerland AG 2019. ISBN 978-3-030-00849-9 (Hardcover), ISBN 978-3-030-00850-5 (eBook).

Cochran, J.K., Mackulak, G.T. and Savory, P.A. (1995). Simulation project characteristics in industrial settings. *Interfaces,* 25(4): 104– 113.

Maria, A. (1997). Introduction to modelling and simulation. pp. 7–13. *In:* Andradóttir, S., Healy, K.J., Withers, D.H. and Nelson, B.L. (eds.). *Proceedings of the Winter Simulation Conference.* http://acqnotes.com/Attachments/White%20Paper%20Introduction%20to%20Modeling%20and%20Simulation%20by%20Anu%20Maria.pdf.

NASA. (2013). *NASA-STANDARDS-7009.* https://spaceflightsystems.grc.nasa.gov/Space DOCII/Standards/documents/NASA-STD-7009.pdf.

NASA. (2016). *NASA-STANDARDS-7009a.* https://standards.nasa.gov/standard/nasa/nasa-std-7009.

Robinson, S. (2008). Conceptual modelling for simulation part I: Definition and requirements. *Journal of the Operational Research Society,* 59(3): 278–290.

Wild, R. (2002). *Operations Management,* 6th Edn., London: Continuum.

# 13

# Modelling of Cyber–Physical–Social Systems*

Chapter 13 presents the basics of Cyber-Physical-Social Systems modelling. To describe the structure of these systems, methods based on multi-agent models have been proposed, with an autonomous agent as the basic element. The basic advantages of multi-agent modelling are described, as well as IT tools to support modelling processes. The following section discusses how to model the processes carried out within the Cyber-Physical-Social Systems. The most widely used process modelling standards are presented, namely: Event-driven Process Chain model, Entity Relationship Diagram, Data Flow Diagram, Petri networks, and Value Stream Mapping. The third part of the chapter presents methods for modelling the risks of loss of process continuity within complex systems of the Cyber-Physical-Social type. A new breakdown of risk sources was proposed from the following perspectives: human resources, financial resources, intangible assets, and infrastructure resources. In the final section of the chapter, a model based on a power distribution is presented as particularly suitable for describing extreme phenomena occurring, for example, in natural disasters.

## 13.1 Modelling of Complex Structures

Typical reliability structures for systems of not-so-high complexity are presented in the Section 6.4 (System Reliability Modelling – the Structural Reliability). These include the following models: (a) serial, (b) parallel, (c) serial-parallel, (d) standby, (e) bridge, and (f) mixed. In addition, hierarchical, tree-based, network-based, and multilevel

---

* https://orcid.org/0000-0002-2630-3507.

models, with or without feedback loops, are also used in the field of control and management. However, all these models do not consider the specifics of complex systems of the type Cyber-Physical-Social ones. Models that do not have these limitations currently appear to be so-called multi-agent models, and the method that uses them is *Agent-based modelling* (ABM).

ABM is a computational framework to represent process dynamics and prepare for simulation experiments (Agent-based simulation – ABS). An *autonomous agent* (AA) can act by pursuing his own goals, if they do not contradict the overarching goals of the entire system. Modelling a population of autonomous agents with its own characteristics and behaviours, which interact with each other and with the environment, is a main feature of an ABM. ABS is usually used to model the process of individual decision-making as well as social and organizational behaviour (Bonabeau, 2001). Agents generally represent people, or groups of people, and their mutual relations symbolize processes of social interaction (Gilbert and Troitzsch, 2005). In a complex system, agents are organizations, which make decisions related to the management of production or services (e.g., material sourcing and ordering, stocking, shipping, capacity expansion, etc.). In an agent-based model composed of artificial agents, collaborating entities communicate their findings to collectively accomplish a task. The development of ABM tools, the availability of data on agent interactions, and advances in computation have made possible a growing number of ABM and ABS applications across a diversity of domains and disciplines.

A typical structure of an agent-based model has three types of elements: agents (their attributes and behaviours), agent relationships and procedures of interaction (an underlying topology of connectedness defines how and with whom agents interact), as well as agents' environment (agents exist in and interact with their environment). Figure 13.1 shows a simplified diagram of how an 'agent' works, and the three basic elements will be discussed in more detail below.

(a) Agents

In the context of agent-based modelling and simulation we consider that agents have following properties and attributes (based on Macal and North, 2010):

- *Autonomy* – an agent is autonomous and self-directed. It can function independently in its environment and in interactions with other agents, generally from a limited range of situations that are of interest and that are included in the model. An agent's behaviour refers to a general process, which is based on monitoring the state of the environment. The selected signals are tracked by sensors and

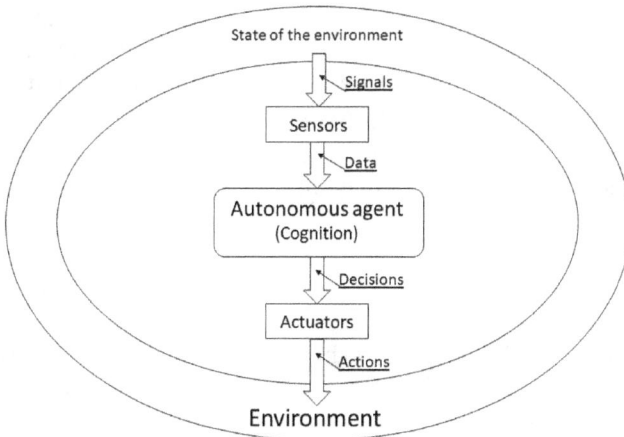

**Figure 13.1:** A simplified diagram of how an 'agent' works.

processed into data, which forms the basis (input) for the autonomous agent's cognitive processes. The result of these processes are decisions passed on to actuators for execution, which in turn process them into actions on specific parts of the environment (see Figure 13.1).

- *Modularity* – autonomous agents are modular or self-contained. An agent is an identifiable, discrete entity with a set of characteristics, behaviours, and decision-making capability. The modularity requirement implies that an agent has a boundary, and it should be possible to determine whether an element is a part of an agent or not.

- *Sociality* – an agent has social type properties, interacting with other agents. Common agent interaction protocols include contention for space and collision avoidance, agent recognition, communication and information exchange, influence, and other application-specific abilities.

- *Conditionality* – the agent's state changes over time and is dynamic in nature. The state represents the system condition, defined by the essential variables associated with its current situation. An agent's state consists of a set of its attributes and its behaviours. The state of an agent-based model is the collective state of all the agents along with the state of the environment. The agent's behaviour is determined by its state.

- *Adaptability* – an agent may be adaptive, e.g., by having rules or instruments that modify its behaviours. Agent behaviour modification can occur because of the learning process, usually based on its collective experiences. Learning and adaptation ability requires having memory as a dynamically updated attribute of the agent. At the system level, learning and adaptation can be modelled by aggregate changes in individual behaviours or by allowing agents to leave the system, increasing the system effectiveness with the more efficient agents.

- *Purposefulness* – an agent can have targets set that drive its behaviour, and criteria by which it can assess the effectiveness of its decision or action. This allows an agent to compare the outcome of its behaviours relative to given goals and adjust its responses in future interactions.

- *Heterogeneity* – agents may be diverse and heterogeneous. Agent characteristics and behaviours may vary in their extent and sophistication, they depend on how much information is considered in the agent's decisions, the agent's internal models of the environment, the agent's view of the possible reactions of other agents in response to its actions, and the extent of memory of previous events the agent retains and can use in making its decisions. Agents may also be equipped with different amounts of resources or accumulate different types of assets because of agent interactions.

(b) Agent relationships and procedures of interaction

The primary issues of modelling agent relationships are specifying who should be connected to whom and describing the dynamics of the interactions.

Common topologies for representing social agent interactions include (Macal and North, 2014):

- 'Soup' – a non-spatial model in which agents have no locational attribute,

- Grid or lattice – cellular automata represent agent interaction patterns and available local information by a grid or lattice. Cells directly surrounding an agent are its neighbourhood. An agent's location is determined by the grid cell index.

- Euclidean space – agents are in 2- or 3-dimensional Euclidean spaces. An agent's position is its relative or geospatial coordinates.

- Geographic Information System (GIS) – agents move over and interact with realistic patches of geo-spatial landscapes. An agent's location is a geographical unit (e.g., a zip code) or geospatial coordinates.

- Networks – network structure can be static (links prespecified) or dynamic (links determined endogenously by relationship-creating mechanisms). An agent's location is the relative node location in the network.

The essential requirement is that agents only interact at any given time with a limited number of other agents out of the population. This notion is implemented by defining a local neighbourhood and limiting interaction to a small number of agents that are near them. Agents do not have to be near one another spatially to be able to interact. The *network topology* allows agents to be linked based on the relationships in addition to proximity. Any agent may be a member of many different populations (e.g., networks).

(c) Agent environment

Agents interact actively with their environment and with other agents. The surroundings may be used to provide information on the spatial location of an agent relative to other agents or it may provide a full set of geographic information (e.g., as in a GIS). An agent's location, included as a dynamic attribute, is needed to track agents as they move across a landscape (e.g., traceability in supply networks), contend for space, acquire resources, and encounter other situations. Complex environmental models can be used to model the agent's surroundings. In some cases, the environment may be seen as a limitation to the agent's operation and constrain its actions. For example, the environment in an agent-based transportation model would include the infrastructure and capacities of the nodes and links of the road network. These capacities would create congestion effects (e.g., reduced speed of movement) and limit the number of agents moving through the transportation network at any given time.

One of the most significant advantages of ABM is the possibility of modelling the phenomenon of *emergence*. Even simple agent-based models in which agents are completely described by deterministic rules and use only local information can self-organize and sustain themselves in ways that have not been programmed into the models. More complex models representing real-world phenomenon can also exhibit emergent behaviour resulting from agent interactions. Agent-based modelling algorithms based on emergence have led to specialized optimization techniques, such as ant colony optimization and particle swarm optimization, that have been used to solve many practical problems (e.g., Bonabeau et al., 1999; Barbati et al., 2011).

Identifying set of agents, accurately and precisely specifying their behaviours, and appropriately representing agent relationship are the keys for developing useful agent models. The first step in the process of developing an agent-based model consists of identifying the agent types according to their attributes. Agents are generally the decision-makers in a system, whether they are human, organizational, or automated. Once the agents are defined, agent behaviours are specified. A normative model in which agents attempt to optimize a well-defined objective can be a useful starting point to developing more descriptive and domain-specific heuristics. You can also start with formalization based on a framework such as the Belief-Desire-Intent (BDI) model (Wooldridge, 2009). To define the agent relationships and procedures of interaction you can use the methods that control Which agents are interacting with Which one, and how they interact.

Modern software toolkits (e.g., based on an object-oriented approach) use a template design approach in which recurring elements are codified and reused for new applications. The Unified Modeling Language (UML) provides a set of tools in the form of diagrams for object-oriented software system design and representation that is independent of computer code implementation (Booch et al., 2005). Some agent-based models include a variety of advanced capabilities such as: machine learning algorithms, geographical information systems (GIS), connections to relational databases, version control, and integrated development environments (IDEs).

ABM can offer distinct advantages to conventional simulation approaches such as discrete event simulation (Law and Kelton, 1991; Law, 2007), system dynamics (Sterman, 2000, 2001) and other quantitative modelling techniques. The agent-based modelling and simulation can be beneficial, if at least one of the following requirements is met:

- The problem has a natural representation as being comprised of agents;
- The decisions and behaviours can be well-defined;
- Agents have behaviours that reflect how individuals and/or entities actually behave;
- Agents can adapt and change their behaviours;
- Agents have the ability to learn and engage in dynamic strategic interactions;
- Agents have dynamic interactions with other agents, and agent relationships form, change, and decay;
- Agents have a spatial component to their behaviors and interactions;
- The structure of the system does not depend entirely on the past, and new structure can emerge that can describe how the system will evolve in the future;
- When process structural change needs to be an endogenous result of the model, rather than an input to the model.

The complete set of object class definitions and methods, parameter values, and initial values for all the agent and other object states constitutes a complete specification of an agent-based model. You can implement such a model by either writing an object-oriented software using programming languages (for example, the Java or C++), or using a higher-level agent-based toolkit such as AnyLogic (AnyLogic, 2018), NetLogo (NetLogo,

2018), Swarm (Bonabeau, 1999; SWORM, 2018), and many others. Any of these toolkits provides an extensive set of classes that encapsulate the basic functionality required by the agent models. For example, the functionality for the sequence of agent operations and interactions in a complex system model and the control mechanisms that cause each of the agent methods to be invoked at the proper time or in the proper situation would be part of the functionality provided by the scheduler class of an agent-based software (Macal and North, 2006).

## 13.2  Process Modelling within Complex Systems

Rational management of processes within complex systems can be divided into four main stages:

- Identification of key processes;
- Modelling of this processes;
- Implementation of processes;
- Introduction of process monitoring and evaluation.

The *identification* of key processes in a complex system is aimed at recognizing those processes that are necessary to achieve the system's or organization's goals. This procedure can take place in two ways:

- 'Top-down' approach – from top to bottom, starting from system's goals or the company's strategy by defining the few key processes, to describing many support processes, and
- 'Bottom-up' approach – from bottom to top, starting from elementary operations, actions, or activities, by joining them into larger groups, up to individual processes and finally, the macro-processes.

As a first step the valuation of processes should be made (e.g., determining the importance of each process). The final effect of this step is to determine: the type, number, content, and structure of processes which are necessary to achieve the organization's objectives and meet the requirements of the clients.

The aim of *process modelling* is the qualitative evaluation and design or modification of processes that are most important to the organization, so that they ensure effective and efficient functioning of the system or organization. In process modelling, two main types of approach are used (based on Bukowski, 2019):

- Diagnostic – starting with the current state (what is the current situation?) with an evolutionary nature (a step-by-step gradual improvement of existing processes), and
- Prognostic – beginning with the desired state (how it should ultimately be?) with an adaptive nature (designing the ideal and striving to achieve the model state).

By modelling a designed process, the following rules should be followed:

- Separation of the process from its environment – each process must have its boundaries, beginning, and an end.
- Process structuring – each process has its own internal structure and consists of sub-processes and/or procedures.
- Determining the responsibility for the process – each process has its 'owner'.
- Defining the subject of the process – each process performs some functions (e.g., production or service).
- Concentration on creating value – the components of the process which do not create added value should be eliminated, or the number of such processes should be kept to a minimum.
- Shaping the course of the process – the best (due to the accepted criteria) configuration should be set for each process.
- Process input security – for each process it is necessary to ensure a reliable supply (e.g., materials, energy, information).

The *implementation* of processes is to create the conditions for the efficient and effective introduction of a new or improved process and ensuring its proper functioning. The scope of activities in this area includes securing assets and resources (e.g., materials, equipment, premises, etc.), and preparing the process owner and process teams to supervise and manage these processes.

The designed and implemented process should be continuously *monitored* and periodically *evaluated* which requires ensuring the following conditions:

- Gathering of measurable parameters of the attributes that uniquely characterize a given process;
- Development of the methodology of measurements (e.g., tools, methods, and accuracy);
- Setting the limit values for individual parameters and signalling methods in case of exceeding them (e.g., warning, alarm).

Control of monitoring results takes place in the following four steps:

1) Continuously or discreetly measurement in real time;
2) Determination of deviations from target values (e.g., required, normative, or recommended);
3) Assessment of deviation significance and analysis of its probable causes;
4) Development of remedial activities to correct results and prevent possible nonconformities in the future.

The frequency of the control depends on the nature of the parameter. The following are typical recommendations for the rate of measurements:

- Process efficiency should be measured and analysed at short intervals (e.g., hourly);
- The duration of the process, timeliness, and quality at medium time intervals (e.g., weekly);
- Process costs at rather long-time intervals (e.g., monthly);
- Customer satisfaction in the longest intervals (e.g., internal quarterly, and external half-yearly, or even annual).

To organize the diversity in process modelling methodologies, several standards have been introduced, and the commonly used are presented below (based on Bukowski, 2019).

a) EPC – Event-driven Process Chain Model

Modelling of event-driven processes enables EPC (Event-driven Process Chain) diagrams that can also illustrate flows in production and service processes. Graphic symbols of EPC diagrams allow to present process control structures in the form of a chain of events and functions (EPC 2018). In practice, patterns are used that allow the user to easily create a graphical model of any event-driven processes (e.g., Microsoft Office Visio EPC Diagram).

The typical EPC diagrams use the following blocks:

- Functions – basic rectangular shaped blocks. Each function corresponds to the performed activity.
- Events – presented in the form of rhombuses, occur before and/or after the function. So, each function can be connected to another function through events.
- Connectors – events and functions are combined with them. There are three types of connectors, corresponding to the following logical operators: AND, OR, and exclusive OR (XOR).

The EPC event chain model can be written as an ordered five:

$$EPC = (E, F, C, m, A) \tag{13.1}$$

where:

$E$ – a non-empty set of events,

$F$ – a non-empty set of functions,

$C$ – a set of connectors,

$m$ – a mapping that assigns the appropriate logical operator to each connector (e.g., AND, OR, XOR),

$A$ – a set of arcs connecting the vertices E, F, and C.

b) ERD – Entity Relationship Diagram

Entity Relationship Diagram is based on three basic categories (ERD, 2018):

- Entity – an object as an element of the reality in which information can be stored. The object can be a living being (e.g., decision-maker, operator, client, supplier, etc.), a thing (e.g., computer, machine, material, product, etc.), an event (e.g., command, contract, forecast, order, etc.) and a place (e.g., plant, department, warehouse, airport, harbor, etc.).

- Relationship – a connection between two or more entities. They can be binary (between two entities, e.g., employee-warehouse) or multiple (between three, four, or more entities).

- Attribute – a feature characterizing entities and relationships in a selective (keys), descriptive (verbal) and procedural (quantitative) manner. Attributes are divided into primary (basic features of a given entity or compound) and derivatives (are determined based on the values of other attributes).

ERD modelling processes takes place in four steps:

- Identification of entities – e.g., by means of SWOT analysis or heuristic methods;

- Development of a preliminary scheme of entity relationships – e.g., using situational analysis;

- Identification of attributes and relationships – e.g., analysing documents and files;

- Developing a full diagram of entity relationships – e.g., based on the rules of system operation and entity matrices.

The method based on entity relationships is structurally mature and is well suited to both analysis (auditing) and synthesis (design) of processes, especially information flows ones. In situations where the processes are complex and their mutual relations very complex, using of ERD requires support by expert teams.

c) DFD – Data Flow Diagram

The DFD is a technique of analysis and design of processes based on a structural approach. It consists of a data flow graph, which describes the required functions of data processing, supplemented with a description of the activities carried out. Diagrams connect process functions to system objects based on an object-oriented methodology. DFD modelling techniques are particularly frequently used as tools for the analysis and design of transaction systems and processes (DFD, 2018).

DFD technique is based on four basic categories:

- Processes – functions transforming input data into output data;
- Data flows – movements from one system element to another;
- Depots – data storage for a specified time, in the form of homogeneous groups; and
- External – objects (so-called 'terminators') that are not part of the analysed system (e.g., customer, supplier, bank, etc.)

Visualization of the method is based on the modification of the symbolism used in graph theory. Process decomposition is usually executed hierarchically 'top-down', starting with a context diagram, and ending with the specification of those elementary functions which are not subject to further decomposition. The degree of process complexity translates into the number of intermediate levels representing the intermediate diagrams (e.g., complex processes are modelled at least on five hierarchical levels). Depending on the number of elements on each level of the hierarchy, DFD diagrams can be symmetrical or asymmetrical. The use of DFD diagrams is particularly useful in the case of processes that are accompanied by intense information flows, such as information flows or distribution logistics. Generally, the inductive thinking and a usage of the process approach dominate in the process of implementing the DFD method.

d) Petri Networks

Petri's network graph is an ordered triplet of the form:

$$N = (P, T, A) \tag{13.2}$$

where:

$P$ – a non-empty set of places (e.g., nodes of the network),
$T$ – a non-empty set of transitions (e.g., material flows),
$A$ – a set of relations between places (e.g., arcs of the network).

The generalized Petri network is an ordered fifth of the form

$$NG = (P, T, A, W, M) \tag{13.3}$$

where:

$(P, T, A)$ – the $N$ network described by equation (13.2)
$W$ – a function of arc weights, which assigns a natural number to each transition as a weight,
$M$ – a function defined on the set of places called the initial marking of the $N$ network. As a result of marking the network, its places are assigned non-negative integers called tags.

Places in the network represent passive system elements and can take certain states and collect certain objects (e.g., goods, information, clients, etc.). *Transitions* are active elements of the system and fulfil the role of transport or processing of different objects. *Relations* represent the structure of the network, showing interconnections of individual network elements and directions of possible transitions. A graphic representation of the Petri network is the graph, in which the places *P* are represented in the form of circles, the transitions *T* in the shape of rectangles, and the relations *A* in the form of directed arcs. Marks in the form of dots are placed inside circles (places), which can be interpreted as the degree of load of a given place or fulfilment of certain conditions. If all entrance places to a given passage have a marker, then this transition is considered active and meets the conditions for the so-called 'rebuff'. As a result of each rebuff, the marking of the network changes, because the entry points to the rebuffed passage lose one marker and the starting points gain one marker.

Petri networks are an effective tool for modelling processes, allowing for conducting any structural analysis and testing of system dynamics with simulation methods. This is an important advantage of the method, especially for complex systems (e.g., networks, or networks of networks) characterized by significant variability (Arnold, 1995).

e)  VSM – Value Stream Mapping

Value Stream Mapping is a technique that allows modelling the flow of values within any production or service process (Seth and Gupta, 2007). One of the main objectives of using this method is to identify and eliminate unnecessary activities in the process and adjust the process to the level of customer orders. The output parameter is a measure that results from the customer's demand for products and determines the rate of implementation of the process subsequent stages. While mapping the value stream one should also consider the striving to minimize the inventory. The value stream map is a flow diagram of information and materials in the process. Value stream mapping consists of three steps:

- Step 1. Value Stream Analysis (VSA) – making a diagnosis of the existing state based on the current value stream analysis.
- Step 2. Value Stream Designing (VSD) – creating a vision of the future state by building the target status of the value stream.
- Step 3. Value Stream Work Plan (VSP) – developing a plan for improvement and implementation of solutions.

The current state map should present:

- The state of the stream on a specific day of analysis;
- The state of the stream for the selected product or service, being a representative of a family of certain products or services;
- Stream condition for average demand;

- Characteristics of key stream providers;
- Characteristics of key clients or groups of customers.

The main purpose of the existing state map is:

- Presentation of the relationship between material and information flows;
- Identifying problems and waste in the stream;
- Determining the time used to process the components into a finished product/services expressed with *L/T* (*Lead Time*), and determination of the cash freezing period that was spent on components;
- Ascertaining the time used to process components into a finished product/services expressed with *P/T* (*Processing Time*), that is, determining the possibility of shortening LT for the stream, by streamlining the flow, eliminating stocks and waiting times;
- Creating the foundations for the construction of a state map for the future system configuration and an action plan.

The general objectives of the mapping process are as follows:

- Build a system for managing the order fulfilment process;
- Define the principles of customer service, planning, supplying, inventory management and flow in the manufacturing process, enabling the execution of orders in the time required by the client and a form adapted to the value perceived by the client;
- Adapt the process to self-improvement and adaptation to a changing environment at the optimal costs;
- Create both short-term (e.g., one year) and long-term (e.g., several years) improvement plan;
- Build a production and service strategy.

In practice, a mapping procedure based on three pillars is recommended: first, perceive the material and information flow, as well as inventories and process limitations, second, understand the transition time, customer needs, and other problems, and finally, improve the whole value stream instead of local improvements. These methods are described in more detail and illustrated with examples by, among others (Bukowski, 2019).

## 13.3 Modelling Process Continuity Disruptions in Complex Systems

A prerequisite for the effective and efficient operation of complex systems is to ensure their continuity in performing the required functions. A threat to the realization of this goal is primarily the risk associated with the possibility of disruption, generally understood as a potential possibility of realizing an undesirable scenario or event associated with the occurrence of negative consequences. The source of risk is a factor or an agent that alone or in combination with other agents has the potential to create possibilities of specific consequences. A *threat* 'refers to the source of risk if its consequences may cause a material damage' (e.g., damage or destruction of infrastructure), while a *hazard* is a 'source of risk whose potential consequences concern safety' (e.g., threats to human life or health).

In specialized literature, many attempts have been made to classify sources of risk occurring in complex systems that are more-or-less universal in its nature. The most general seems to be the classification recommended by the Federal Emergency Management Agency (FEMA P-789, 2013), according to which sources of risk are divided into three main categories, i.e., external, internal, and process-related ones. Based on this classification, it is proposed the following division of potential risk sources:

I. External risk sources

- Explosions – nuclear attack or detonation, radiological attack (e.g., dirty bomb), and explosives attack;
- Biological attack – aerosol anthrax, plague, ricin, food contamination, and animal diseases (e.g., foot and mouth disease);
- Pandemic influenza;
- Chemical attack or accident – blister agent, nerve agent (paralysing the nervous system), toxic industrial chemicals, and chlorine tank explosion;
- Infrastructure damage (critical infrastructure attack or failure) – power failure (blackout), communication system failure or disruption, water supply contamination, sewage system failure, heating, ventilation and air conditioning systems failure, and major fire;
- Cyber-attack – loss of data, and computer network outage;
- Economic and social – economic catastrophe (market crash, loss of trust), demonstrations, riots and civil unrest, labour dispute, and mass transit strike;
- Natural disasters – high winds (hurricane, tornado), winter storm, thermal anomalies, major earthquake, flood, tsunami, and volcano eruption.

II. Internal risk sources

- Sabotage;
- Planning errors and mistakes;

- Computer system faults and failures;
- Lack of competent service;
- Bad atmosphere within the organization or crew dissatisfaction.

III. Process risk sources

- Insufficient critical supply or lack of adequate stock;
- Lack of trustworthy partners and/or suppliers;
- Improper process configuration;
- Unreliability of critical system components or single points of failure.

The above-stated risk sources are related to the causes of possible threats and hazards that concern the basic resources of the organization, namely, physical, cyber, financial, human, and social. This perspective, called the *resource approach*, can be the basis for classifying threats and hazards also in cyber-physical-social systems. Based on the literature and our own works, the following division of threats in relation to cyber-physical-social systems is proposed:

A. Perspective of human resources

a) External threats:
- explosive (e.g., nuclear, radiological, bomb attacks),
- biological attacks (e.g., diseases, poisoned food and water, epidemics, and pandemics),
- chemical attacks and catastrophes (e.g., toxic, burning, explosive, paralysing agents),
- social related (e.g., strikes, anxieties, riots, demonstrations).

b) Internal threats:
- sabotage,
- planning mistakes,
- lack of competence.

B. The perspective of financial resources

a) External threats related to:
- market (e.g., dependence of prices on the market),
- currency (e.g., exchange rate changes),
- interest rate (e.g., changes in interest rates on the market),
- purchasing power (e.g., the impact of inflation on the purchasing power of the currency),
- politics (e.g., resulting from political decisions).

b) Internal threats related to:
- finance (e.g., resulting from liabilities to foreign capital),
- business (e.g., related to the changeability of income earned),
- financial liquidity (e.g., related to the fulfilment of contractors' obligations),
- transaction (e.g., non-compliance with the terms of the transaction).

C. Perspective of intangible assets
- theft of know-how,
- loss of key information.

D. Perspective of infrastructure resources
a) External threats:
- power network failure,
- transport system failure,
- water supply failure,
- telecommunications system failure.

b) Internal threats
- damage to the internal infrastructure (e.g., storage and warehousing facilities, transshipment terminals, ports, logistics centres),
- computer system failure.

All these risks are random in nature, and as such can be described using the models presented in Section 6.2. Particularly useful are the probability distributions derived from the assumptions of random Poisson processes, for which the relevant formulas for evaluating their parameters and describing the shape of the distribution are collected in Table 6.1. Probability distributions that are not derived from these random processes can be also used to model rare events, and their maximum values. These are mostly distributions based on the extreme value theory as well as those that relate to the distribution of rank data frequencies. *Extreme values distributions* are founded on statistics built on time series of random samples taken at identical time intervals, in long time intervals, in which a random sample is formed by extreme values from a given time interval (maximum or minimum result). In the case of exposure modelling, the maximum values characterizing the given parameter are particularly important, therefore the Gumbel distribution can be a useful model, if observations or measurements are available from a sufficiently long-time interval (Gumbel, 1941). It should be highlighted that the expected value of the extreme increases with the length of the time segment it concerns. A classic example of the practical and effective use of this model to predict possible floods in Black Canyon on the Colorado River is the work (Gumbel, 1941) described in detail.

A particularly useful distribution for modelling unevenly distributed data is the power distribution. It is based on the assumptions of the so-called Zipf's Law, which refers to frequency distributions of rank data. Originally, Zipf's law stated that, in natural language, the frequency of any word is roughly inversely proportional to its rank in the frequency table. Thus, the most frequent word will occur approximately twice as often as the second most frequent word, which occurs twice as often as the fourth most frequent word, etc. (Hawa, 2007). The *power distribution* was created as a generalization of Pareto diagrams and the Zipf's law. Georg Kingsley Zipf stated that the distribution of the frequency of individual words in different languages is subject to the simplest form of power distribution with an exponent equal to –1. In later studies, this distribution has proved particularly useful in the practice of modelling single random events, such as natural disasters (e.g., earthquakes, forest fires, floods etc.). General model in the simplest form can be presented as following:

$$Y = a \, x^b \tag{13.4}$$

which after nonlinear transformation can be presented in the form of a linear relationship

$$y' = c + bx' \tag{13.5}$$

where: $y' = log \, y$;   $x' = log \, x$;   $c = log \, a$

where the purpose of modelling is to describe a value that describes the magnitude of an event, the exponent $b$ must have a negative value. A graphical example of dependencies described by the power model with the exponent b = –0.4 is shown in Figure 13.2, in form of the distribution of rainfall frequency as a function of their intensity.

To summarize the above considerations, it can be concluded that the prediction of exposures with the nature of single, repetitive events is currently not a problem that is difficult to solve, and the software tools supporting the quantitative assessment of random variables parameters are widely available. The *chains of events* leading to the loss of the continuity of processes could theoretically be modelled by simple probabilistic dependencies or using stochastic processes, in particular the so-called Markov and Semi-Markov processes. These models have been described in detail in the literature (Eberle, 2015), therefore they will not be discussed in detail in this chapter. However, in practice, most of the assumptions underlying these models are not met, so the usability of these type of models is in many cases inadequate.

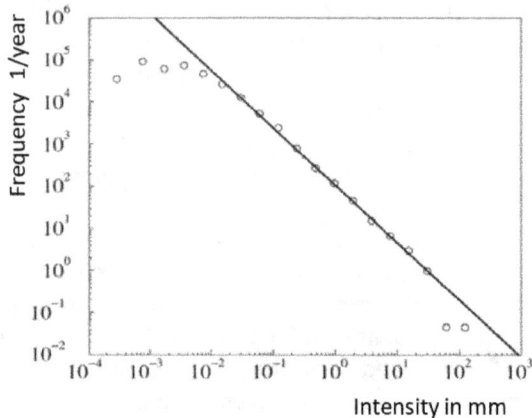

**Figure 13.2:** Distribution of rainfall frequency as a function of their intensity (Sornette, 2009).

## 13.4 References

AnyLogic. (2018). http://www.anylogic.com/features.

Arnold, D. (1995). *Materialflusslehre*. Braunschweig, Wiesbaden: Vieweg.

Barbati, M., Bruno, G. and Genovese, A. (2011). Applications of agent-based models for optimization problems: A literature review. *Expert Systems with Applic.*, 39(5): 6020–6028.

Bonabeau, E. (2001). Agent-based modeling: Methods and techniques for simulating human systems. *Proceedings of the National Academy of Sciences*, 99(3): 7280–7287.

Bonabeau, E., Dorigo, M. and Theraulaz, G. (1999). *Swarm Intelligence: From Natural to Artificial Systems*. Oxford: Oxford University Press.

Booch, G., Jacobson, I. and Rumbaugh, J. (2005). *The Unified Modeling Language User Guide*, 2nd Edn. Boston: Addison-Wesley.

Bukowski, L. (2019). *Reliable, Secure and Resilient Logistics Networks. Delivering Products in a Risky Environment*, Springer Nature Switzerland AG 2019. ISBN 978-3-030-00849-9 (Hardcover), ISBN 978-3-030-00850-5 (eBook).

DFD. (2018). https://www.lucidchart.com/blog/data-flow-diagram-tutorial.

Eberle, A. (2015). *Markov Processes*. https://wt.iam.uni-bonn.de/fileadmin/WT/Inhalt/ people/ Andreas_Eberle/MarkovProcesses/MPSkript1415.pdf.

EPC. (2018). https://www.ariscommunity.com/event-driven-process-chain.

ERD. (2018). https://www.lucidchart.com/pages/er-diagrams.

FEMA P-789. (2013). *Continuity Guidance Circular 2. Continuity Guidance for Non-Federal Governments: Mission Essential Functions Identification Process (States, Territories, Tribes, and Local Government Jurisdictions)*. https://ready.alaska.gov/Plans/continuity/ Continuity-Guidance-Circular2.pdf.

Gilbert, N. and Troitzsch, K. (2005). *Simulation for the Social Scientist*. McGraw-Hill. 2nd Edn. UK: McGraw-Hill Education.

GMU (George Mason University). 2014. MASON Home Page. http://cs.gmu.edu/~eclab/ projects/mason/.

Gumbel, E.J. (1941). The return period of flood flow. *Ann. Math. Statistics*, 12(June): 163–190.

Hawa, M. (2007). *Probability Distribution Summary*. https://www.google.com/search? Ei=vnNTW-TNFsPKwAL217G4BQ&q.

Law, A.M. (2007). *Simulation Modeling and Analysis*. 4th Edn. New York: McGraw-Hill.

Law, A.M. and Kelton, W.D. (1991). *Simulation Modelling and Analysis*. NY: McGraw-Hill, Inc.

Macal, C. and North, M. (2006). Tutorial on agent-based modeling and simulation part 2: How to model with age. pp. 73–83. *In*: Perrone, L.F., Wieland, F.P., Liu, J., Lawson, B.G., Nicol, D.M. and Fujimoto, R.M. (eds.). *Proceedings of the 2006 Winter Simulation Conference.*

Macal, C. and North, M. (2010). Tutorial on agent-based modeling and simulation. *Journal of Simulation*, 4(3): 151–162.

Macal, C. and North, M. (2014). Introductory tutorial: Agent-based modeling and simulation. pp. 6–20. *In*: Tolk, A., Diallo, S.Y., Ryzhov, I.O., Yilmaz, L., Buckley, S. and Miller, J.A. (eds.). *Proceedings of the 2014 Winter Simulation Conference.*

NetLogo. (2006). *NetLogo Home Page.* http://ccl.no rthwestern.edu/netlogo.

Seth, D. and Gupta, V. (2007). Application of value stream mapping for lean operations and cycle time reduction: An Indian case study. *Production Planning & Control*, 16(1), 1 January 2005: 44–59.

Sornette, D. (2009). Dragon-kings, Black swans, and the prediction of crises. *Int. Journal of Terraspace Science and Engineering.* http://www.arxiv.org›physics.

Sterman, J.D. (2000). *Business Dynamics: Systems Thinking and Modeling for a Complex World.* Boston: Irwin McGraw-Hill.

Sterman, J.D. (2001). System dynamics modeling: Tools for learning in a complex world. *California Management Review*, 43(4) Summer: 8–25.

SWARM. (2018). *The Swarm Simulation System, a Toolkit for Building Multi-agent Simulations.* http://www .santafe.edu/projects/swarm/overview /overview.html.

Wooldridge, M. (2009). *An Introduction to Multiagent Systems.* 2nd Edn. John Wiley & Sons Ltd.

# 14

# Simulation of Cyber–Physical–Social Systems Behaviour in a Risky Environment[*]

The main purpose of modelling systems and processes, especially complex ones of the Cyber-Physical-Social type, is to create opportunities for simulation studies. Chapter 14 presents the basics of the methodology for building simulation models and using them to generate various cases that have not yet been observed in practice but may occur in the future. Particularly, this applies to hazardous events that pose a threat to the environment, human safety, and the continuity of production or service processes. The first section discusses the principles of the two basic methods on which simulation models are based, namely System Dynamics Modelling and Simulation (SDMS) and Discrete Event Simulation (DES). The second section is devoted to a practical example of simulating a complex system behaviour under uncertainty. The main objective of this example is to demonstrate the feasibility of using the proposed framework to assess the vulnerability and resilience of global supply networks. The subject of the research was a commodity steel plant located in Central Europe, and the main goal of the research was to determine the recommended actions of decision-makers (modelled using a multi-agent technique) in the occurrence of disruptions in the supply of raw materials necessary for steel production.

[*] https://orcid.org/0000-0002-2630-3507.

## 14.1 Basic Principles of Complex Systems Simulation

In practice, the following methods are most often used to simulate complex Cyber-Physical-Social systems: System Dynamics Modelling and Simulation (SDMS) and Discrete Event Simulation (DES). SDMS offers a methodology to support the analysis of dynamic processes and captures the factors affecting the behaviour of the whole system under consideration in a causal-loop diagram. This diagram clearly represents the linkages and feedback loops among the individual elements in the system, as well as all relevant linkages between the system and its environment. This type of analysis can be especially useful to a decision-maker in understanding a complex engineered system by creating a variety of scenarios and observe how the system might perform under different conditions (Sweetser, 2009).

*System Dynamics* (SD) is a computer-aided approach for analysing and solving complex problems and especially the dynamics of changing their behaviour over time. This area of research was initiated by Jay W. Forrester at the Massachusetts Institute of Technology, and was initially called Industrial Dynamics (Forrester, 1958). System Dynamics has its roots in control engineering and management science, and the approach uses a perspective based on information feedback as well as delays in the system's response to stimuli. SD is a universal approach and allows you to understand the dynamic behaviour of complex cyber, physical, biological, and social systems. Forrester (1961) defines Industrial Dynamics as

**... the study of the information-feedback characteristics of industrial activity to show how organizational structure, amplification (in policies), and time delays (in decision and actions) interact to influence the success of the enterprise. It treats the interactions between the flows of information, money, orders, materials, personnel, and capital equipment in a company, an industry, or a national economy.**

System Dynamics is well appropriate specially to modelling continuous processes, systems where behaviour changes in a non-linear fashion, and systems where extensive feedback occurs within the system. SD models often incorporate qualitative aspects of behaviour that might significantly affect the performance of a system. Its causal loop diagrams are an effective way of describing feedback and linkages within a system. However, animation associated with a running System Dynamics Modelling and Simulation is usually limited to updating graphs and numerical displays (Morecroft, 2015). In this concept, the components, and relationships among the components of a system build the structure of the system which determines its behaviour. Based on the proper defining of the linkages between people, organizations, processes, and resources, the structure of a system can be optimized to improve its performance. These links are modelled by feedback loops, where a change in one variable affects other variables in the system, which makes the models time-varying (that is, dynamic). Another crucial concept in system dynamics is the so-called 'mental model'. These types of models can be characterized as flexible, informative, and manage to integrate data from diverse sources. A significant part of the SDMS effort is therefore associated with capturing these mental models using a causal loop diagram that represents the system's dynamic behaviour.

To map the dynamic behaviour of a complex system, one needs to identify and describe all feedbacks relevant to the system, both positive (+) and negative (–) (Gharajedaghi, 2006). Figure 14.1 shows an example of building an SD-type model for the process of changing the abundance of a certain population of living organisms in three steps. The method combines both qualitative and quantitative aspects to explore, realize, and communicate complex and sophisticated problems (Forester, 1985; Sterman, 2000). The qualitative part entails the creation of *Causal Loop Diagrams* (CLD), as illustrated in Figure 14.1a, b in which variables are drawn in a cause-and-effect relationship pattern, which creates the hypothesized dynamic structure of the system. A cause-and-effect relationship can either change the behaviour in the same direction – i.e., increase (indicated using a plus sign) or in the opposite direction – i.e., reduce (indicated using a minus sign). A loop with a B indicates that it is

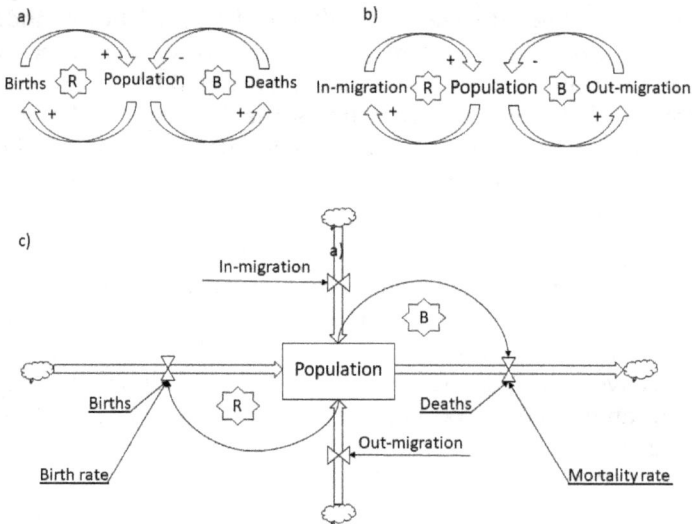

**Figure 14.1:** An example of building an SD-type model for the process of changing the abundance of a certain population of living organisms.

balanced and seeking stability in the system, and a loop with a R stands for reinforcing (Richardson, 1986). The quantitative aspect involves the development of a process model based upon a *Stock and Flow Diagram* (SFD) as illustrated in Figure 14.1c, and mathematical equations which represent interrelated variables in the system. Stock variables (rectangles) represents the state variables and are the accumulations in the system. Flow variables (valves) alter the stocks by filling or draining the stocks. The arrows point the causal relation between two variables and reflect the flow of information within the model structure (Sterman, 2001). Such models can be used to experiment with alternative scenarios by changing the values of the variables in the model which, when run, produce output (a simulated dynamic behaviour). The output of the simulation can then be compared with the real world and provide answers to 'what if' questions (Richmond, 2001). The level of detail in the model is related to the project questions, which create the boundaries of the model and determine which variables should be included. Additionally, all variables are defined explicitly, which makes the procedure transparent since all variables and their functions are observable.

SDMS models the behaviour of systems using differential equations. Because of the nature of these mathematical functions, the SDMS is particularly useful for modelling continuous systems. However, SDMS is less well suited to providing a detailed representation of a system with discrete changes in state variables, or mixed systems of both discrete and continuous processes (e.g., inter-arrival rates of discrete entities in a system). The application of SDMS to logistics and supply chain management has its origins in the fundamental works of Forrester (1958, 1961), in which his models of a production-distribution systems are described in terms of six interacting flow systems, namely: the flows of information, materials, orders, money, manpower, and capital equipment. Based on the development and use of a System Dynamics simulation model, Forrester describes, analyses, and explains issues evolving around supply chain management (Angerhofer, 2000).

The term *Discrete Event Simulation* (DES) is used for the modelling approach based on the concept of entities, resources, and block charts describing entity flow and resource sharing (Cassandras and Lafortune, 2009; Fishman, 2001). This approach has its origins to the 1960s when Geoffrey Gordon conceived and evolved the idea for GPSS and brought about its IBM implementations (Gordon, 1961). General Purpose Simulation System (GPSS) is a discrete time simulation general-purpose programming language, where a simulation clock advances in discrete steps. Entities (or transactions in GPSS) are passive objects that represent people, parts, documents, tasks, messages, etc. They flow through the blocks of the flowchart where they wait in queues, are delayed, processed, split, combined, etc. (Borshchev and Filippov, 2004). A properly built DES model can determine the performance of an existing system very closely and provide a decision-maker insights into how that system might perform if modified, or how a completely new system might perform. To achieve a satisfactory trustworthiness to the performance of a real process, a DES model requires precise and correct data on how the system operated in the past, or accurate estimates on the operating characteristics of a proposed system. Like SDMS, DES also gives the decision-maker the ability to model and compare the performance of systems and processes for different scenarios and alternatives. DES is more appropriate to the detailed analysis of a specific, well-defined system or linear process, such as a manufacturing line or a service centre. These systems can change at specific points in time in various ways, such as resources fail, operators take breaks, shifts change, or infrastructure damage occurs. DES can also provide statistically valid estimates of performance measures associated with these systems or processes, such as number of entities waiting in a particular queue or the longest waiting time for service.

Most software based on the Discrete Event Simulations concept include graphs and numerical displays, as well as a computer animation of the system. In these animations, icons represent entities moving through a graphical representation of the system. The process flow visualization in a DES-animation can be a particularly useful tool to improve understanding of a process. Thus, it is used to gain an understanding of how an existing system behaves, and how it might behave if given changes are made to the system. Some examples of application of the above-described modelling and simulation methods are included in the works (Bukowski and Karkula, 2003, 2009, 2010).

## 14.2  A Practical Example of Simulating a Complex System Behaviour Under Uncertainty

One of the main problems in complex logistics systems, especially within the global supply network, is the disruption risk and loss of supply continuity (Security Service, 2006; Waters, 2007). It results from the efforts to minimize costs by introducing Lean Manufacturing, Lean Logistics, and Lean Management as well as from the increasing number and intensity of external threats. Reducing the number of suppliers according to the 'Four S' principle (a Single Source Supply Strategy), introducing to a greater extent the 'Just in Time' manufacturing system, minimizing the level of buffers, and configuring tightly connected supply chains have resulted in a significant increase in the level of delivery disruption risk. Based on the main trends in the risk understanding (e.g., Aven, 2014, 2020; Aven et al., 2014) by the author's team (e.g., Bukowski et al., 2015, 2016) has been proposed the framework for disruption risk modelling in the complex logistic networks.

The main objective of this example is to demonstrate the feasibility of using the framework to assess the vulnerability and resilience of complex logistics systems with the nature of global supply networks. The essential idea of the overall concept is based on the experiences gained in the supply chain risk management, dependability engineering, as well as in the field of the resilience engineering. Based on the rules of Samson's Unified Service Theory (Sampson, 2010), a general model of a complex logistics system is proposed, consisting of many parts representing both the inward and the outward logistics (Figure 14.2). The left part of the model represents upstream operations, related to supply of the focus organization in raw materials, necessary to carry out the production. The right part of the model includes downstream operations and processes corresponding to the distribution of products to the first- and second-tier customers.

The model is dynamic and considers the imaginable disruptive events, e.g., disruption in supply process and unforeseen changes in demand. The structure of this model is based on Agent-based modelling (see Section 13.1 as well as: D'Inverno and Luck, 2004; Swaminathan et al., 1998). Implementation of the model has been made on the example of steel mills, whose logistics supply and demand system has a global scope. Simulation studies on this model were designed to evaluate the vulnerability and resilience of the whole logistics system to possible disturbances.

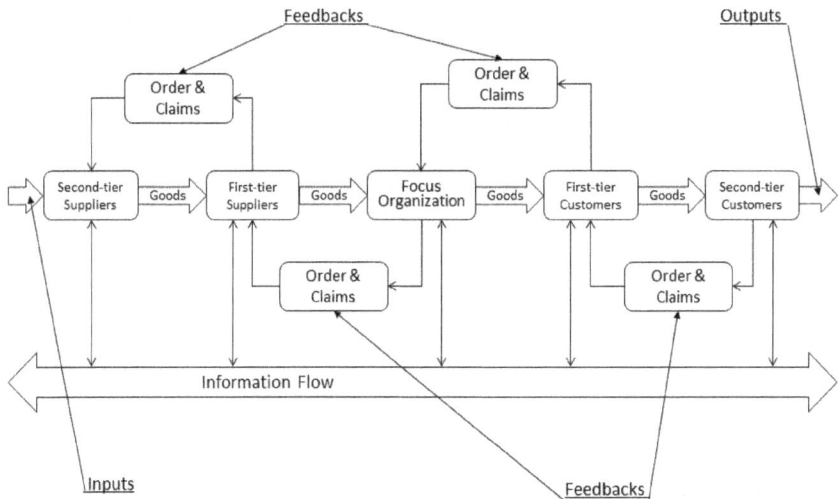

**Figure 14.2:** General model of a complex logistics system, consisting of parts representing both the inward and the outward logistics.

a)  Logistic networks vulnerability and resilience metrics

Even though the ideas of vulnerability and resilience were introduced relatively recently, they have already been surveyed in many serious studies, both theoretical and practical areas of engineered systems (e.g., AMBER, 2009; Aven, 2011; Dekker et al., 2008; Hollnagel et al., 2006; Kröger and Zio, 2011; Sanchis et al., 2012; Sheffi, 2007). Moreover, in logistics and supply chains a significant number of interesting works has been published (e.g., Cavinato, 2004; Christopher and Peck, 2004; Jüttner, 2005; Kleindorfer and Saad, 2005; Malone, 2006; Natarajarathinam et al., 2009; Peck, 2003, 2006; Sheffi and Rice, 2005; Stemmler, 2006; Svensson, 2000; Waters, 2007). A common feature of these works is the concept that can be called *process continuity-oriented approach* which has been the subject of numerous works (e.g., Bukowski, 2014, 2016; Bukowski and Feliks, 2012, 2014; Bukowski et al., 2015, 2016, 2017), in which it were proposed the following basic definitions of the key terms:

- *Continuity* – a system capability to deliver products or services at an acceptable, predefined performance level under the real work conditions (e.g., even with disruptive events).
- *Disruptive event* (DE) – an act of delaying or interrupting the process continuity (e.g., system failure, natural catastrophe, manmade fault).
- *Vulnerability to a disruptive event (V)* – the degree to which a system is affected by a disruptive event.
- *Vulnerability metric* – the disruption impact described by two main indicators: expected loss of performance ($L_D$), and disruption time ($T_D$).
- *Resilience to a disruptive event* (RES) – the ability of a system to absorb and withstand the disruption impact, and still continue to deliver products or services at an acceptable predefined performance level, as well as the adapt-capacity to new work conditions.
- *Resilience metric* – the collective term described by the three main indicators: absorbability (ABS), recoverability (REC), and adaptability (ADA).

Based on these assumptions and the generalized functional resilience model (see Section 9.2 and Figure 9.2), the specific model of a Typical course of a product delivery process with a continuity disruption is proposed as shown in Figure 14.3. The thick line shows the course of an idealized system operation as it changes its performance in function of time. Before the disruption occurs, the system was functioning at the required level of performance $P_{Req}$. The occurrence of a disruptive event is immediately followed by a sharp decline in system performance (in the figure it reaches a minimum level of performance higher than a critical

PERFORMANCE - P

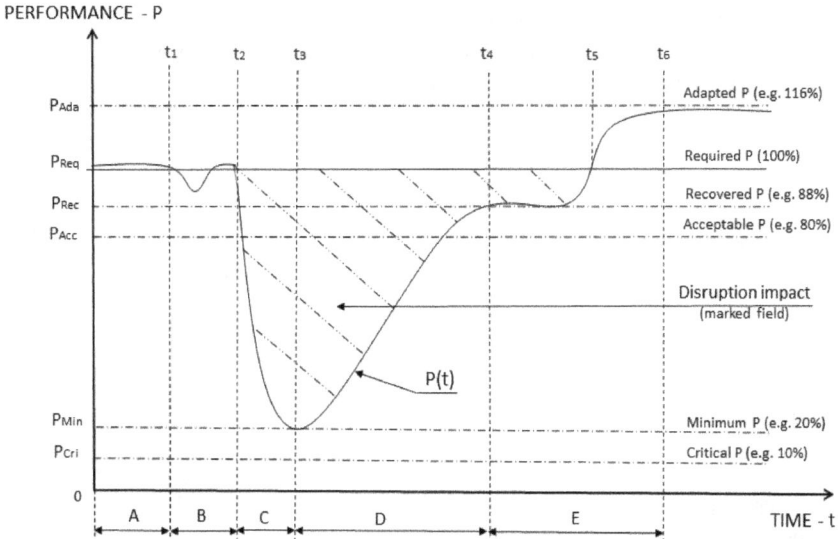

**Figure 14.3:** Typical course of a product delivery process with a continuity disruption.

level of performance: $P_{Min} > P_{Cri}$). With the capacity to absorb the effects of a disruptive event, the system maintains its basic functions and gradually increases its performance. After the time $t_4$, it achieves a recovered performance level, which lies above the acceptable level of performance ($P_{Rec} > P_{Acc}$). The final phase of the course is characterized by the ability to adapt to new conditions, and as a result the system performance improves (optionally).

Usually, there can be distinguished the following five fundamental phases in a typical course of a service delivery process with a continuity disruption (Figure 14.3):

Resistant state A – characterized by no reaction to small disturbances,

Robust behaviour B – with short-term loss of performance after a disturbance and rapid return to the required state,

Absorption phase C – distinguished by 'coping' with disruption and continuity retain of operation,

Recovery phase D – characterized by 'bounce back' to the acceptable performance level,

Adaptation phase E – distinguished by 'learning' from disruption and transformation to the new work conditions.

The vulnerability and resilience properties appear in the C, D, and E phases, so the disruption curve shape will be the basis for a quantitative assessment of these properties. The loss of performance is proportional to the area between the line showing the required performance $P_{Req}$ and the actual course of the performance (disruption impact $L_D$ shown as marked field). Therefore, the general metric for vulnerability can be described by the two-dimensional variable (duplet):

$$V = < L_D, T_D >$$ (14.1)

where:

disruption impact $\qquad L_D = \int_{t2}^{t5} [P_{Req} - P(t)]dt$ (14.2)

disruption time $\qquad T_D = t_5 - t_2$ (14.3)

$P_{Req}$ – required performance level,

$P(t)$ – performance at the time ,

$t_2$ – beginning of disruption,

$t_5$ – end of disruption.

Resilience metrics is represented as a collective term described by the three-dimensional vector as follows:

$$RES = < ABS, REC, ADA >$$ (14.4)

where:

$$ABS = \frac{P_{Min}}{P_{Req}}$$ (14.5)

$$REC = \frac{P_{REC}}{P_{Req}}$$ (14.6)

$$ADA = \frac{P_{ADA}}{P_{Req}}$$ (14.7)

b) Research subject

The subject of the research was a commodity steel plant located in Central Europe (Bukowski et al., 2015). The mill is supplied with large amount of raw material required for a production, of which the basic one is an iron ore (in the amount of approximately 5.5 million tonnes per year). The materials for a production process are being delivered to the stockpile buffer, which is characterized by a limited storage capacity, which determines the maximum possible level of the reserve volume. Due to the significant storage costs of resources, the stock of raw materials has been set at a level that ensures the continuity of the mill's production under normal and uninterrupted supply conditions. Iron ore is imported from East Europe (railway transport) and Brazil (combined transport). In the investigated company, iron ore is mainly delivered from Ukraine. From Ukraine ore is transported to Poland directly by railway carriages.

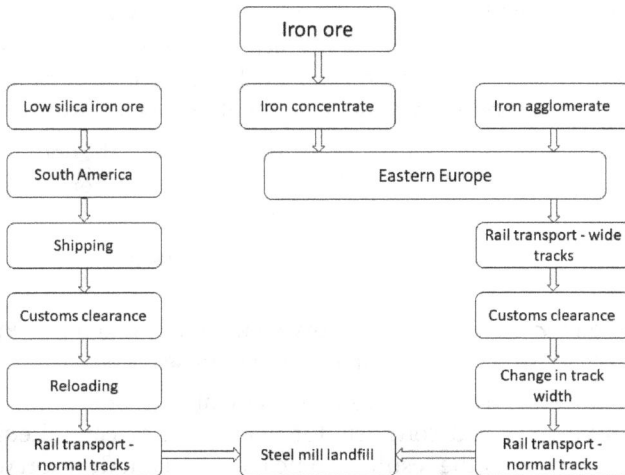

**Figure 14.4:** Simplified scheme of supply of key raw materials for steel production at the studied steel mill.

Depending on the distance between the mine and Polish border as well as the weather conditions, the delivery time, i.e., the entry of a train into the steelworks area varies from 2 to 5 days. The delivery can be realized in different ways depending on the current conditions. Typically, supplier A (Ukraine) covers 80% of the demand for raw materials, while the remaining 20% is provided by supplier B (Brazil). In case of supply disruptions from the above-mentioned suppliers, the demand for raw materials may be covered by the delivery from a supplier C (Serbia), but this involves higher transport costs and longer time of delivery. The simplified scheme of supply of key raw materials for steel production at the steel mill under consideration is shown in Figure 14.4. The timeliness of deliveries depends on many factors which are often unpredictable. For example, the transport processes can be disrupted by breakdowns in the means of transport, or (in winter) inability to discharge the cargo due to its freezing. Such cases cause the discontinuity of supplies whose effects may be different, including the production stoppage.

c) Research methodology

To investigate the effect of disturbances on the continuity of production process, the simplified simulation model of the raw materials supply network was constructed. The demo version of the software AnyLogic (www.anylogic.com) was used, which allows to connect different techniques of the system modelling. These techniques include modelling of the complex system structure with Agent-based modelling (ABM) – see Section 13.1, and simulation of continuous processes using the methods of SDM and DES – see Section 14.1.

The basic elements used to describe the behaviour of the system are stocks, flows, and information. Typical business activities include at least one of the following five types of stocks: materials, personal, capital equipment, orders, and money, which are defined as follows:

- Materials includes all stocks and flows of physical goods which are part of a production and distribution process, whether raw materials, in-process inventories, or finished products.
- Personnel generally refers to people or hours of labour (e.g., man-hours).
- Capital equipment includes factory space, tools, and other equipment necessary to produce goods and delivery of services.
- Orders include such things as procurement of goods, requisitions for new employees, and contracts for new space or capital equipment. Orders are usually the result of some management decision which has been made, but not yet converted into the desired result.
- Money is used in the cash sense, which means that a flow of money is the actual transmittal of payments between different stocks of money.

Stocks define static part of the system. Flows describe how values of stocks change in time and thus define the dynamics of the system.

The developed model combines all mentioned above ways of modelling. The main part of the model, reflecting the realization of supply chain and shown schematically in the Figure 14.3, was built from the elements used in the modelling of dynamic processes. The model was built to account for all possible disruptions in the timely delivery of materials. The risks arising from the disruption within a supply chain and the ways of preventing their adverse effects were modelled using agent-based systems, while moments in which disturbances occur were modelled as discrete random events using appropriate probability distributions. The main task of the agents is an appropriate response to the presence of disruption within the supply chain to sustain continuity of a production by providing the required number of raw materials to the plant. The expert team identified 15 of the most important scenarios for disruptive events which have been investigated using computer simulation (Table 14.1). Moreover, the following typical options were analysed for each scenario:

I. Lean strategy – 'Just in Time' supplying with no resources in the stockpile.
II. Sustainable safe strategy – resources in the stockpile covering the five-days production.
III. Crisis strategy – resources in the stockpile covering the ten-days production.

**Table 14.1:** The scenarios for disruptive events.

| Scenario | Before Disruption (% of Delivery) | | | After Disruption (% of Delivery) | | |
|---|---|---|---|---|---|---|
| | Sup. A | Sup. B | Sup. C | Sup. A | Sup. B | Sup. C |
| S1 | 80% | 20% | none | none | 20% | 80% |
| S2 | 80% | 20% | none | none | 100% | none |
| S3 | 80% | 20% | none | none | 40% | 60% |
| S4 | 80% | 20% | none | none | 20% | none |
| S5 | 60% | 40% | none | none | 40% | 60% |
| S6 | 60% | 40% | none | none | 100% | none |
| S7 | 60% | 40% | none | none | 60% | 40% |
| S8 | 60% | 40% | none | none | 40% | none |
| S9 | 50% | 50% | none | none | 50% | 50% |
| S10 | 50% | 50% | none | none | 100% | none |
| S11 | 50% | 50% | none | none | 70% | 30% |
| S12 | 50% | 50% | none | none | 50% | none |
| S13 | 20% | 80% | none | none | 80% | 20% |
| S14 | 20% | 80% | none | none | 100% | none |
| S15 | 20% | 80% | none | none | 80% | none |

In the absence of any disturbance, the realization of option I is the cheapest, but in the situation of a major supply disruption, it leads to a complete stoppage of steel production, which results in huge losses.

The interruption in the supply of raw material may be:

a. short – less than 5 days,
b. average – 6 to 10 days, or
c. long – more than 11 days.

The average delivery time for each supplier was: supplier A – 2 days, supplier B – 10 days and supplier C – 3 days. In all cases the constant demand is assumed. Based on the model shown in Figure 14.5, 1,000 simulation experiments for each case were carried out. The results of these simulations are shown in Table 14.2 providing the percentage production loss for specific scenarios designated for the quarter (3 months) of a year, as a practical measure of the system vulnerability to the disruptive events.

d) Research results

For the well visualization of the production losses, it was assumed that during the 'normal' operation the production level corresponds to the level of the orders, while during the crisis the level of production may

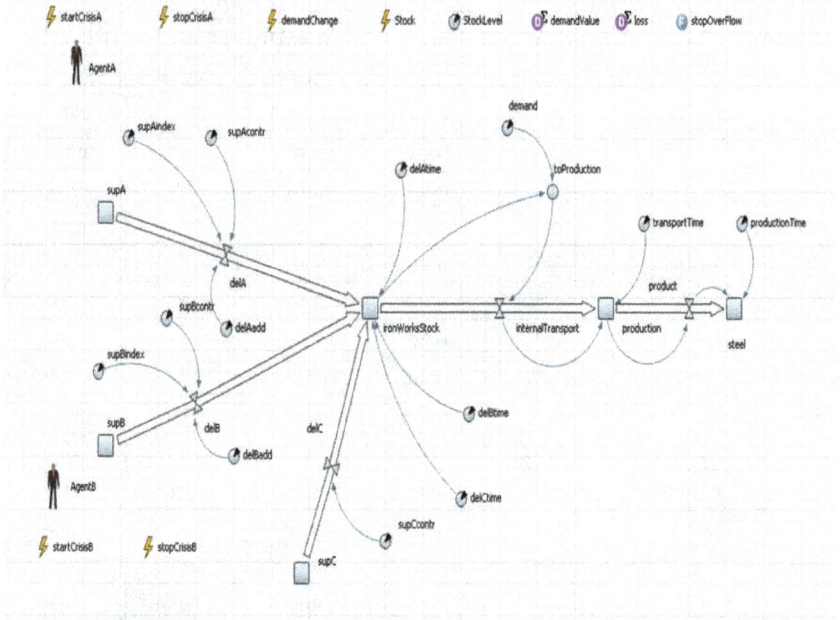

**Figure 14.5:** Simulation model of the logistics network supply system for the steel mill under study.

**Table 14.2:** The percentage of production loss for each scenario.

| Scenario | I a | I b | I c | II a | II b | II c | III a | III b | III c |
|----------|-----|-----|-----|------|------|------|-------|-------|-------|
| S1 | 3,56 | 5,92 | 5,92 | 0 | 0,39 | 0,39 | 0 | 0 | 0 |
| S2 | 3,56 | 7,98 | 11,78 | 0 | 2,43 | 6,23 | 0 | 0 | 0,74 |
| S3 | 3,56 | 6,43 | **7,30** | 0 | 0,88 | 1,75 | 0 | 0 | 0 |
| S4 | 3,56 | 8,00 | 12,46 | 0 | 2,45 | 6,90 | 0 | 0 | 1,34 |
| S5 | 2,67 | 3,99 | 3,99 | 0 | 0 | 0 | 0 | 0 | 0 |
| S6 | 2,67 | 5,97 | 8,56 | 0 | 0,43 | 3,01 | 0 | 0 | 0 |
| S7 | 2,67 | 4,64 | 5,47 | 0 | 0 | 0 | 0 | 0 | 0 |
| S8 | 2,67 | 6,00 | 9,33 | 0 | 0,45 | 3,78 | 0 | 0 | 0 |
| S9 | 2,22 | 3,32 | 3,32 | 0 | 0 | 0 | 0 | 0 | 0 |
| S10 | 2,22 | 4,98 | 7,10 | 0 | 0 | 1,55 | 0 | 0 | 0 |
| S11 | 2,22 | 3,98 | 4,83 | 0 | 0 | 0 | 0 | 0 | 0 |
| S12 | 2,22 | 5,00 | 7,78 | 0 | 0 | 2,22 | 0 | 0 | 0 |
| S13 | 0,89 | 1,33 | 1,33 | 0 | 0 | 0 | 0 | 0 | 0 |
| S14 | 0,89 | 1,99 | 2,77 | 0 | 0 | 0 | 0 | 0 | 0 |
| S15 | 0,89 | 1,99 | 2,89 | 0 | 0 | 0 | 0 | 0 | 0 |

be lower, due to the lack of raw materials. Below, there is presented the exemplary scenario description and the results of a simulation for the case in which the break in supplies A lasts 14 days. The operation of the agent system aiming to maintain production 'sets off' additional deliveries in the proportion of supplier B – 40% and supplier C – 60% of the demand for raw materials. This scenario, shown in Figure 14.6a, illustrates an adverse situation because it additionally assumes the lack of any reserve on the stockpile (scenario S3-Ic). After 70 days, a break in delivering from supplier A begins (black continuous line). In the fifth day of the crisis, the delivery from the Supplier B (green 'triangle' line) is increased from 20% to 40% and the delivery from the Supplier C (yellow 'point' line) starts in the amount of 60% of the total demand for raw materials. Extra deliveries provide 100% production demand during the crisis and return to 'normal' levels (supplier A – 80%, supplier B – 20%, supplier C – 0%) after the crisis stops.

In Figure 14.6b, the level of production and demand for a completed product is presented. As a result of the crisis the production drops down to about 20% (in the beginning of the crisis only regular deliveries from Supplier B are being provided), however, after several days the production

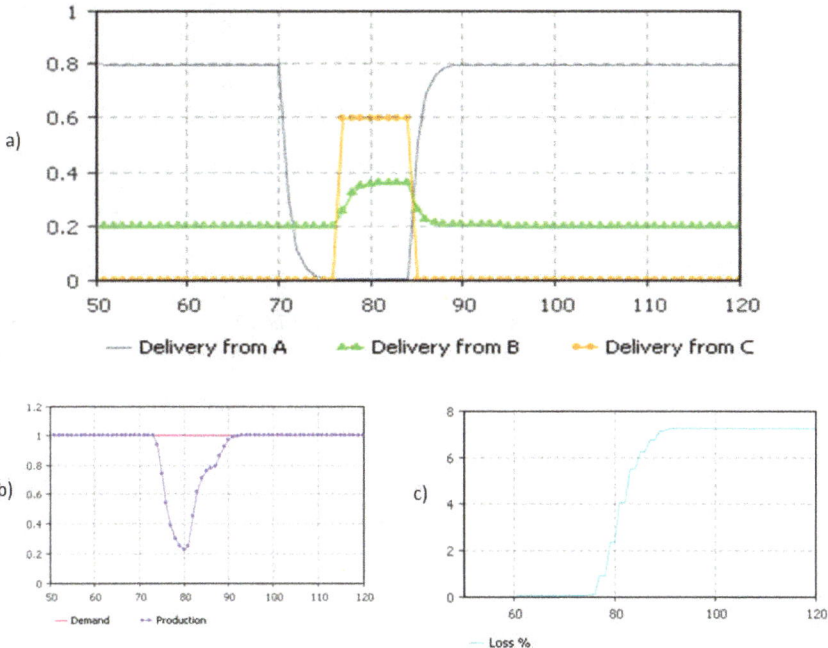

Figure 14.6: Simulation results for the scenario S3-Ic: a) the percentages of supplies for A, B, C suppliers; b) Production and demand level; c) the percentage loss of production calculated per quarter.

level rises – first up to the level of 80% (after about 8 days, extra supplies from supplier C arrived) and then, after about 15 days, to the level of 100% (increased delivery from Supplier B also enters the steel mill).

In Figure 5.9c, the percentage loss of production calculated per quarter is shown. The losses appear around a day of 75 and grow for about 15 days. The total production loss in relation to the demand for a completed product is 7.3%. Thus, the vulnerability measure of the tested system in this case (scenario S3-Ic) reaches the value V = < 7.3%; 15 days >.

The similar simulation tests were carried out for all scenarios and for all possible combinations I, II, III, and a, b, c. The results of these tests are shown in Table 14.2 in the shortened form and may serve as a preliminary assessment of the tested system vulnerability to threats resulting from different risky scenarios. According to the assumptions presented in this model the agent actions are restricted, and oriented primarily at preventing production losses, or if that is not possible then reduce to a minimum.

## 14.3 References

AMBER. (2009). *Assessing, Measuring, and Benchmarking Resilience*, FP7 – 216295.

Angerhofer, B.J. and Angelide, M.C. (2000). System dynamics modelling in supply chain management: Research review. pp. 342–351. *In*: Joines, J.A., Barton, R.R., Kang, K. and Fishwick, P.A. (eds.). *Proceedings of the 2000 Winter Simulation Conference.*

AnyLogic Simulation Software. http://www.anylogic.com/features, (accessed 15.07.2022).

Aven, T. (2011). On some recent definitions and analysis frameworks for risk, vulnerability, and resilience. *Risk Analysis*, 31(4): 515–522.

Aven, T. (2014). A new perspective on how to understand, assess and manage risk and the unforeseen. *Reliability Engineering and System Safety*, 121: 1–10.

Aven, T. (2020). *The Science of Risk Analysis. Foundation and Practice.* London and New York: Routledge, Taylor, and Francis Group.

Aven, T., Baraldi, P., Flage, R. and Zio, E. (2014). *Uncertainty in Risk Assessment*, Wiley.

Borshchev, A. and Filippov, A. (2004). From system dynamics and discrete event to practical agent based modeling: Reasons, techniques, tools. *The 22nd International Conference of the System Dynamics Society*, 25–29 July 2004, Oxford, England. https://www.systemdynamics.org/assets/conferences/2004/SDS_2004/PAPERS/381BORSH.pdf.

Bukowski, L. (2014). Managing disruption risks in the global supply networks—A transdisciplinary approach. *Proceedings of International Conference on Industrial Logistics*, Croatia, pp. 101–106.

Bukowski, L. (2016). System of systems dependability: Theoretical models and applications examples. *Reliability Engineering & System Safety*, 151: 76–92.

Bukowski, L. (2019). *Reliable, Secure and Resilient Logistics Networks. Delivering Products in a Risky Environment.* Springer Nature Switzerland AG 2019, ISBN 978-3-030-00849-9 (Hardcover), ISBN 978-3-030-00850-5 (eBook).

Bukowski, L. and Feliks, J. (2012). Multi-dimensional concept of supply chain resilience. *Proceedings of the 2nd Carpathian Logistics Congress*, Jesienik.

Bukowski, L. and Feliks, J. (2014). A unified model of systems dependability and process continuity for complex supply chains. pp. 2395–2403. *In*: Nowakowski, T. (ed.). *Safety and Reliability: Methodology and Applications*. London: Taylor & Francis Group, A Balkema Book.

Bukowski, L. and Karkula, M. (2003). The simulation of logistic processes using DOSIMIS-3 simulator. *In: Proceedings of the International Conference on Logistics*, Ostrava 2003, pp. 296–300.

Bukowski, L. and Karkula, M. (2009). Modelling and simulation of logistics processes in heat and power plants: A hybrid approach. *In: Proceedings of the Twentieth International Conference on Systems Engineering – ICSE 2009*. ISBN 978-1-84600-0294; Coventry, United Kingdom.

Bukowski, L. and Karkula, M. (2010). Integration of discrete event simulation, decision tables and fuzzy systems in modelling of logistics processes. pp. 113–121. *In*: Swiatek et al. (eds.). *Advances in Systems Science* (ICSS 2013)' Academic Publishing House EXIT.

Bukowski, L., Feliks, J. and Majewska, K. (2015). Modelling and simulation of disruption risk in the complex logistic networks: A multimethod approach. *In: Safety and Reliability of Complex Engineered Systems*. London: Taylor & Francis Group, A Balkema Book, pp. 3911–3918.

Bukowski, L., Feliks, J. and Majewska, K. (2016). Logistic system resilience modelling: A dynamic, multiagent, service engineering-oriented approach. *In: Risk, Reliability and Safety: Innovating Theory and Practice*. London: Taylor & Francis Group, A Balkema Book, pp. 2207–2214.

Bukowski, L., Feliks, J. and Majewska, K. (2017). A modeling framework for the resilience analysis of steel mill logistic supplying systems. *Proceedings of the 25th International Conference on Metallurgy and Materials "Metal 2016"*, Brno 25-27.05.2016, 2017, 1613–1619, ISSN 978-80-87294-67-3.

Cassandras, C.G. and Lafortune, S. (2009). *Introduction to Discrete Event Systems*. Berlin: Springer-Verlag.

Cavinato, J.L. (2004). Supply chain logistics risks. *International Journal of Physical Distribution and Logistics Management*, 34(5): 383–387.

D'Inverno, M. and Luck, M. (2004). *Understanding Agent Systems*. Berlin: Springer.

Dekker, S. et al. (2008). *Resilience Engineering: New directions for Measuring and Maintaining Safety in Complex Systems*. Final Report December 2008.

Fishman, G.S. (2001). *Discrete-Event Simulation: Modeling, Programming, and Analysis*. Berlin: Springer-Verlag.

Forrester, J.W. (1958). Industrial Dynamics: A major breakthrough for decision makers. *Harvard Business Review*, 36(4): 37–66.

Forrester, J.W. (1961). *Industrial Dynamics*. Portland (OR): Productivity Press.

Forrester, J.W. (1985). The model versus a modeling 'process'. *System Dynamics Review*, 1: 133–134.

Gharajedaghi, J. (2006). *System Thinking. Managing Chaos and Complexity*, Elsevier.

Gordon, G. (1961). A general purpose systems simulation program. *Proceedings of EJCC*, Washington D.C., NY: McMillan, pp. 87–104.

Hollnagel, E., Woods, D.W. and Leveson, N. (eds.). (2006). *Resilience Engineering: Concepts and Precepts*. UK: Ashgate.

https://gistbok.ucgis.org/bok-topics/simulation-modeling.

Jüttner, U. (2005). Supply chain risk management. *International Journal of Logistics Management*, 16(1): 120–141.

Kleindorfer, P.R. and Saad, G.H. (2005). Managing disruption risks in supply chains. *Production and Operations Management*, 14(1): 53–68.

Kröger, W. and Zio, E. (2011). *Vulnerable Systems*. London: Springer-Verlag.

Malone, R. (2006). *Growing Supply Chain Risks*. www.Forbes.com.

Morecroft, J.D.W. (2015). *Strategic Modelling and Business Dynamics. A Feedback Systems Approach*. NJ, USA: John Wiley & Sons Ltd.

Natarajarathinam, M. et al. (2009). Managing supply chains in times of crisis: A review of literature and insights. *International Journal of Physical Distribution and Logistics Management*, 39(2009): 535–573.

Peck, H. (2003). *Creating Resilient Supply Chains: A Practical Guide*. School of Management, Cranfield University, Cranfield, Bedford.

Peck, H. (2006). *Supply Chain Vulnerability, Risk and Resilience, in Global Logistics*, 5th Edn., (Ed.) Waters, D. London: Kogan Page.

Richardson, G.P. (1986). Problems with causal-loop diagrams. *System Dynamics Review*, 2: 150–70.

Richmond, B. (2001). *An Introduction to Systems Thinking, (ithink software)*. High Performance Systems, Inc.

Sampson, S.E. (2010). A unified service theory. *In*: Salvendy, G. and Karwowski, W. (eds.). *Introduction to Service Engineering*. New Jersey: Wiley.

Sanchis, R., Poler, R. and Lario, F.C. (2012). Identification and analysis of disruptions: The first step to understand and measure Enterprise Resilience. *6th International Conference on Industrial Engineering and Industrial Management*, Vigo, 18–20 July.

Security Service. (2006). *Security Advice: Business Continuity*, London: HMSO. www.mi5.gov.uk.

Sheffi, Y. (2007). *The Resilient Enterprise: Overcoming Vulnerability for Competitive Advantage*. MIT Press.

Sheffi, Y. and Rice, Jr. J.B. (2005). A supply chain view of the resilient enterprise. *MIT Sloan Management Rev.*, 47: 41–48.

Stemmler, L. (2006). *Risk in the Supply Chain, in Global Logistics*, 5th Edn., (Ed.) Waters, D. London: Kogan Page.

Sterman, J. (2000). *Business Dynamics – System Thinking and Modeling for a Complex World*. Massachusetts: McGraw-Hill Higher Education.

Sterman, J.D. (2001). System dynamics modeling: Tools for learning in a complex world. *California Management Review*, 43(4) Summer: 8–25.

Svensson, G. (2000). Conceptual framework for the analysis of vulnerability in supply chains. *International Journal of Physical Distribution and Logistics Management*, 30: 731–749.

Swaminathan, J., Smith, S. and Sadeh, N. (1998) Modelling supply chain dynamics: A multiagent approach. *Decision Sciences*, 29(3): 607–632.

Sweetser, A. (2009). *A Comparison of System Dynamics (SD) and Discrete Event Simulation*. https://pdfs.semanticscholar.org/bca5/0943f66fd012dd62168433d5f04221b5d6f5.pdf.

Waters, D. (2007). *Supply Chain Risk Management: Vulnerability and Resilience in Logistics*. London & Philadelphia: Kogan Page Limited.

Part IV

# Managing Risks in Cyber-Physical-Social Systems Under Deep Uncertainty

# 15

# Uncertainty-oriented Concepts of Decision-making[*]

Decision-making under conditions of uncertainty is an important and responsible process, especially if it concerns the continuity of operation of complex Cyber-Physical-Social Systems type structures. Chapter 15 presents the concept of risk due to the imperfect knowledge available to the decision-maker under conditions of uncertainty. Decisions can be made about either taking or not taking given activities that will take place in an uncertain environment. The value of risk depends on the strength of knowledge about the object of the decision available to the decision-maker, the uncertainty associated with the possibility of a given outcome, and the value of consequences of a given outcome. The second part of the chapter deals with the concept of *risk-informed decision-making* (RIDM) as a purposeful process that uses a set of performance measures, together with other considerations, to inform the decision-maker about the risks associated with the consequences of the various decision options. The third section of the chapter proposes the concept of dependably related decision-making in the form of a 10-step algorithm. This concept forms the basis of a generalized cognitive dependability model and Cognitive Dependability Governance Framework for Cyber-Physical-Social Systems operating under deep uncertainty presented in Chapter 16.

## 15.1 Imperfect Knowledge-based Concept of Risk

The concept of risk is ambiguous and covers many areas of human activity. In the common sense, the risk is understood as a situation resulting in

[*] https://orcid.org/0000-0002-2630-3507.

exposure to potential danger. Civil law defines the risk as "the danger of causing harm to the injured". At the beginning of the 20th century, it was supposed that the risk is dependent on subjective uncertainty, and F.H. Knight (1921) published the theory of uncertainty, dividing the uncertainty into measurable (i.e., risk) and non-measurable (i.e., uncertainty *sensu stricto*). Based on these assumptions in the following years, many concepts of risk and their modifications were created, and used for risk assessment by decision-making, especially in business area.

These efforts to unify the concept of risk were summarized by Terje Aven in his work (Aven, 2012), in which a critical review of the risk literature was made, and on this basis the classification of risk definitions into nine main groups proposed. Some of these definitions have only historical value, therefore they have been omitted from further consideration. The others, on the other hand, were divided into three main classes and characterized below (Bukowski, 2019):

a) Risk as expected value of a possible result

This is historically the oldest quantitative approach to risk, attributed to mathematician, A. de Moivre (de Moivre 1738), and still commonly used as a part of financial analysis, as well as by using the so-called decision trees. In this approach, the measure of risk is the product of the occurrence probability of a given event and the financial consequences of this event. This class of definitions may also include the concept of J. Adams (Adams, 1995), according to which the risk can be described by the product of the occurrence probability of a certain event and its utility (e.g., usefulness if it occurs).

b) Risk as a combination of the effects of a certain scenario and the likelihood of this scenario's occurrence

In this class of definitions, the most universal one is the proposal of S. Kaplan and B.J. Garrick (1981), that define the risk as the triplet $(s_i, p_i, c_i)$, where $s_i$ is the $i$-th unwanted disruption scenario, $p_i$ is the uncertainty metric (probability) of this scenario, and $c_i$ is the severity metric (consequence) of the $i$-th scenario, for $i = 1, 2, 3, ... n$. This class includes also the definition recommended by the EN-N-18 002 standard, according to which the risk is a combination of the occurrence frequency or probability of a specific event causing the threat, and consequences related to this event.

c) Risk because of uncertainty in achieving the assumed goals

The most general approach, recommended by the ISO 31000 standard in the 2009 version, based on the idea already proposed by C.O. Hardy (1923). He understood the concept of risk as the uncertainty about the costs, losses, and damage that may occur in an unpredictable future. This approach allows to clarify the 'operational' concept of risk, depending on the specificity of the problem and the perspective from which it is considered. The class c) could also include new trends in risk research, represented for example by Committee on Foundations of Risk Analysis, and presented in SRA Glossary (2018) in the following definition: "Risk is the potential of gaining or losing something of value (such as physical health, social status, or financial wealth) resulting as an outcome from a given activity (planned or not planned) taken in spite of uncertainty."

Based on the above considerations and considering the possibility of uncertainty modelling, it is proposed to provide a universal approach to risk, from a qualitative and quantitative perspective as follows:

We consider a decision-making process and define risk in relation to uncertain future consequences of the decision made. Decisions can be made about either taking or not taking given activities that will take place in an uncertain environment (e.g., natural phenomena). The value of risk (risk metric) depends on:

- the strength of knowledge about the object of the decision available to the decision-maker (e.g., confidence or maturity level),
- the uncertainty associated with the possibility (e.g., probability) of a given outcome, and
- the value of consequences of a given outcome (e.g., number of victims or economic losses).

Risk description for a given *activity A* can therefore be represented as the triplet $(C, U, K)$, where $C$ is some specified *consequences*, $U$ the *uncertainty* associated with $C$, and $K$ the *background knowledge* that supports $C$ and $U$ (which includes a judgement of the strength of this knowledge). The basic terms of this definition can be described as follows:

*Activity (A)* – an intentionally designed and implemented action. We divide activities into operations, tasks, and decisions.

*Consequences (C)* – the effects of an activity with respect to the values defined (such as human life and health, environment, and economic assets), covering the totality of states, events, barriers, and outcomes, and often seen in relation to some reference values (planned values, objectives, etc.)

*Uncertainty (U)* – a situation of having *imperfect knowledge* about the true value of a quantity or the future consequences of an activity, scenario, or event.

*Background knowledge (K)* – the ability to evaluate available information and understand reality in accordance with the current state of knowledge, including a judgement of the strength of this knowledge.

The risk description model can be shown in the following form:

$$R(A) = (C, U, K) \qquad (15.1)$$

where: $A = \{a_i\};$   $C = \{c_i\};$   $U = \{u_i\};$   $K = \{k_i\}$   for $i = 1, 2, 3, ..., n$

In this approach, all three above distinguished classes of risk definitions, namely, a), b), and c), can be described by one, universal model expressed by formula (15.1). In the case of the traditional approach, class a) – the set of consequences can be described in the language of classical sets theory, and the set of uncertainty measures estimated based on probability theory and mathematical statistics. In the most general case, the set of consequences could be described in the language of fuzzy sets theory, and the set of uncertainty measures estimated based on the possibility theory (e.g., possibility and necessity functions).

Based on the model (15.1) a discussion of the role that background knowledge plays in risk analysis can be made. The term *risk analysis* is understood in broad perspective, as a systematic process to comprehend the nature of risk and to express the risk, with the available knowledge (Aven, 2020). It includes risk assessment, risk characterization, risk communication, risk evaluation, risk framing, risk governance, and risk management, as well as policy relating to risk, in the context of risks to individuals, the public, organizations, and to society. Therefore, the comprehensive risk analysis process consists of the following stages (Bukowski, 2019):

- *Risk assessment* – systematic process to comprehend the nature of risk, express, and evaluate risk, with the available knowledge.
- *Risk characterization* – a qualitative and/or quantitative description of the risk, i.e., a structured statement of risk usually containing the following elements: risk sources, causes, events, consequences, uncertainty measurements and the knowledge that the judgments are based on.
- *Risk communication* – exchange or sharing of risk-related data, information, and knowledge between and among different target groups (such as regulators, stakeholders, consumers, media, public).
- *Risk evaluation* – process of comparing the result of risk analysis against risk (and often benefit) criteria to determine the significance and acceptability of the risk.
- *Risk framing* – the initial assessment of a risk problem, clarifying issues and defining the scope of subsequent work.
- *Risk governance* – the application of governance principles to the identification, assessment, management, and communication of risk. Governance refers to the actions, processes, traditions, and

institutions by which authority is exercised and decisions are taken and implemented.

- *Risk management* – activities to handle risk such as prevention, mitigation, adaptation or sharing. It often includes trade-offs between costs and benefits of risk reduction and choice of a level of tolerable risk.

The methodology for conducting risk analysis at all these stages depends primarily on the risk analyst's or decision-maker's knowledge of the risks associated with the problem under consideration. The level of this knowledge (or rather, lack of knowledge) can be assessed in various ways, the most useful of which in practice seems to be the so-called knowledge matrix based on Rumsfeld's concept. Former US Defence Secretary Donald H. Rumsfeld, during a briefing at the Pentagon in 2002, said (Rumsfeld, 2011):

**Reports that say that something hasn't happened are always interesting to me, because as we know, there are known knowns; there are things we know we know. We also know there are known unknowns; that is, to say we know there are some things we do not know. But there are also unknown unknowns – the ones we don't know we don't know.**

This sentence clearly expresses the differences between the *explicit knowledge* (i.e., all existing, available, and accessible knowledge) and the *tacit knowledge* (i.e., so-called hidden meta-knowledge, namely, our knowledge about knowledge). If these relationships are to be represented in the form of a full matrix of the type 2 times 2, it should also include a fourth case, namely 'unknown knowns', i.e., existing knowledge which we are not aware of.

Figure 15.1 shows the two-dimensional knowledge matrix built based on the four possible combinations. In the top right quadrant I, lies the area of *Certainty* (i.e., 'known knowns'), where the extent of what is known is fully understood, and all possible changes are well described. In this area, practically, the risk is negligibly small, including variability subjected to known rules, such as seasonal changes and long-term trends. Below square I lies the zone III of *Unconscious Uncertainty* (i.e., 'unknown knowns'), representing a hidden knowledge to decision-makers. Typically, this is due to the so-called organizational amnesia, which is the result of incorrect knowledge management in a company (e.g., low level of knowledge culture). The top left corner II belongs to *Conscious Uncertainty* (i.e., 'known unknowns'), where the decision-maker is aware of the imperfection of his knowledge. This is the area of typical risk, including variability subjected to random rules. And quadrant IV in the bottom left section represents *Deep Uncertainty* (i.e., 'unknown unknowns'), with pure ignorance of the situation faced (a total lack of knowledge).

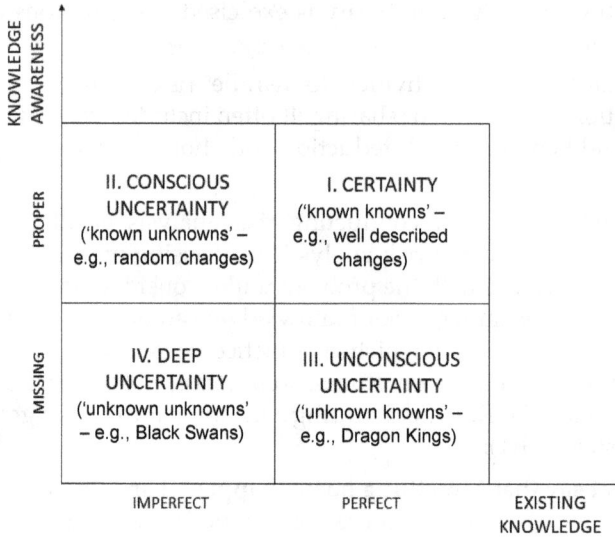

**Figure 15.1:** The two-dimensional knowledge matrix (based on Bukowski, 2019).

Different risk assessment methods are recommended for each of these four areas, namely:

*I. Certainty* – complete perfect knowledge. The background knowledge should be exploited, making full use of known facts and rules to make well-founded decisions and activities. The basic tools used in this area are operational research methods and analytical models.

*II. Conscious Uncertainty* – strong imperfect knowledge. Areas of 'known unknowns' should be thoroughly examined and carefully examined, to understand the causes of uncertainty and the scarcity of needed knowledge. The basic tools used in this area are probabilistic methods and mathematical statistics.

*III. Unconscious Uncertainty* – weak imperfect knowledge. The organizational knowledge culture and management as well as amnesia needs to be exposed and uncovered, e.g., through a facilitated process, to unlock the knowledge that exists and to use it effectively. This area includes also the so-called 'Dragon Kings'.

*IV. Deep Uncertainty* – lack of knowledge. Ignorance can only be tackled through new experiments or by gaining new knowledge using cognitive processes and expertise-based intuition. The main goal of activities in this area should be to strive for growing in both knowledge and its awareness, to reduce the size of this zone, sometimes called 'Black Swan'.

*Dragon King* is a metaphor for an event that is both extremely large in size/impact (a 'king') and belongs to unique origins (a 'dragon') relative to other events from the same system. Such types of events are generated by different phenomena such as positive feedbacks, tipping points, bifurcations and phase transitions that tend to occur in complex nonlinear systems and serve to amplify these events to extreme levels. Seeking to understand and monitor the dynamics of these systems, some predictability of such events may be obtained. The theory of Dragon kings has been developed by Didier Sornette (2009), who hypothesizes that many of the crises' situations that we face may be predictable to some degree. This theory urges that special attention should be given to the study of the system dynamic and monitoring of extremes. From a practical viewpoint, such extremes are interesting because they may reveal underlying, often hidden, organizing behavioural principles. The theory of Dragon Kings is closely related to such concepts as outliers, nonlinear dynamics, power low, extreme value theory, etc.

*Black Swan* can be considered as a metaphor for an event that is surprising (to the observer or decision-maker), has a major effect (e.g., financial impact), and, after being observed, is rationalized in hindsight. The theory of Black Swans is epistemological, relating to the lack or limited knowledge and understanding of the observer. The term was introduced and popularized by Nassim Taleb (2010) and has been associated with concepts such as heavy tails, non-linear payoffs, model error, and 'unknowable unknown' event terminology. Decision-making under deep uncertainty will be further discussed in Section 16.1.

## 15.2 Main Principles of Risk-informed Decision-making Based on Imperfect Knowledge

The concept of a *risk-based decision-making* process provides a defendable basis for decisions and helps to identify the major risks as well as prioritize efforts to minimize or eliminate them. This concept is based primarily on a sparse collection of model-based risk metrics and does not provide many opportunities to interpret these indicators. In contrast, the concept of *risk-informed decision-making* (RIDM) is a purposeful process that uses a set of performance measures, together with other considerations, to 'inform' decision-making (Zio and Pedroni, 2012). The RIDM approach assumes that *human judgement* has a relevant role in decisions, thus technical or economic information cannot be the unique basis for decision-making. This is because of the usually available imperfect knowledge and, also because decision-making process is an intrinsically subjective, value-based procedure. When solving complex problems containing various, competing objectives, the cumulative knowledge provided by experienced personnel is essential for integrating technical and nontechnical elements to produce *dependable decisions* (NASA, 2008, 2010).

RIDM is invoked for key decisions (e.g., design decisions, make-buy decisions, budget reallocation, and many others), which typically require setting or re-baselining of requirements. It is applicable for decisions that typically have one or more of the following characteristics (NASA, 2010):

- high financial stakes – significant costs and important potential safety impacts are involved in the decision;
- complexity – the actual consequences of alternatives are difficult to understand without detailed analysis;
- presence of uncertainty – uncertainty in key inputs creates significant ambiguity in the outcome of the decision alternatives and points to risks that should be managed;
- multiple objectives – large numbers of purposes require detailed formal analyses;
- diversity of stakeholders – high accuracy is needed to define objectives and derive the corresponding performance measures when the set of stakeholders represents an extensive variety of preferences and perspectives.

The risk-informed decision-making is a structured process that (NASA, 2008):

- aims at achieving 'project' success by risk-informing the selection of decision alternatives;
- ensures that decisions between competing alternatives are taken with an awareness of the risks associated with each, thus helping to avoid late design changes, which can be relevant sources of risk, cost overshoot, schedule delays, and cancellation;
- tackles some of the following issues:
  - the possible 'incongruence' between stakeholder expectations and the resources required to address the risks to achieve those expectations;
  - possible misunderstanding of the risk that a decision-maker is accepting when making commitments to stakeholders;
  - the miscommunication in considering the respective risks associated with competing alternatives.
- tries to foster development of a robust technical basis for decision-making by:
  - coupling the attributes of the proposed decision alternatives to the objectives that define 'project' success;
  - considering all attributes that are important to the stakeholders in an integrated manner;

o helping ensure that a broad spectrum of decision alternatives is considered;

o performing quantitative assessment of the advantages and drawbacks of each decision alternative relative to the identified objectives;

o taking into account the uncertainties related to each proposed decision alternative to quantify their impact on the achievement of the identified objectives;

o communicating the quantitative assessment of the proposed decision alternatives into the decision environment, where it is deliberated along with other considerations to form a comprehensive, risk-informed basis for alternative selection.

Within an organizational hierarchy, top-level objectives, such as mission success, flow down in the form of progressively more detailed performance requirements, whose achievement guarantees that the top-level objectives can be met. Each operational or organizational unit within organizational hierarchy discusses with the units at the next lower level in the hierarchy a series of objectives, performance measures and requirements, resources, and schedules that characterize the tasks to be performed by this units. At each step, the lower-level unit manages its own threats and reports risks and elevates decisions for managing risks to the next higher level. Throughout this process, interactions take place between the following actors (NASA, 2008):

- the stakeholders (i.e., individuals or organizations that can be affected by the outcome of a decision, but they are not parts of them);
- the risk analysts (i.e., individuals or organizations that apply uncertainty-based methods to the quantification of risks and performances);
- the subject matter experts (i.e., individuals or organizations with expertise in one or more areas within the decision domain of interest);
- the technical authorities;
- the decision-maker.

The analyses are performed by subject matter experts in the domains spanned by the objectives. The completed risk analyses are deliberated, and the decision-maker selects a decision alternative for implementation with the agreement of the technical authorities. The NASA RIDM process consists of three parts summarized below and described in detail (Dezfuli et al., 2010; NASA, 2008, 2010):

Part I – Identification of decision alternatives

Main objectives are decomposed into their sub-objectives, each of which reflects an individual issue that is significant to the stakeholders. At the lowest level of decomposition are measures of performance objectives, each of which is associated with a performance measure that quantifies the degree to which the performance objective is addressed by a given decision alternative. In general, a performance measure indicates the direction of increasingly good performance measure values. A complete set of performance measures is considered for decision-making, that reflects stakeholder interests and spans the mission execution domains of safety, technical, cost and schedule (NASA, 2010). Objectives whose performance measure values are required to lie within predefined boundaries, i.e., in the range between the minimum and maximum values, give rise to imposed constraints reflecting those limits. Objectives and imposed constraints constitute the basis on which decision alternatives are compiled, and performance measures are how their ability to meet obligatory constraints and achieve objectives is quantified.

Part II – Risk analysis of decision alternatives

The performance measures of each alternative are quantified. In the presence of uncertainty, the actual outcome of a particular decision alternative should be only one. Therefore, it is obligatory on risk analysts to model all possible outcomes of interest, accounting for their probabilities or possibilities of occurrence, in terms of the scenarios that produce it. If the uncertainty in one or more performance measures avoids the decision-maker from assessing important differences between alternatives, then the risk analysis should be iterated to reduce uncertainty. The iterative analysis stops when the level of uncertainty does not preclude a robust decision from being taken. A *robust decision* is based on sufficient evidence and characterization of uncertainties to determine that the selected alternative best reflects the decision-maker's preferences and values at the current level of knowledge available to the decision-maker and is considered resistant to credible modelling perturbations and realistically foreseeable new information (NASA, 2010). The principal product of the risk analysis is the Technical Basis for Deliberation (TBfD), a document that lists the set of possible alternatives, summarizes the analysis methodologies used to quantify the performance measures, and presents the results. The TBfD is the input that risk-informs the deliberations supporting the decision-making process.

Part III – Risk-informed alternative selection

The process of deliberation takes place among the stakeholders and the decision-maker either reduces the set size of alternatives and requests for further analysis of the remaining alternatives, or chooses an alternative for implementation, or inquires for new alternatives. To simplify deliberation, a set of performance requirements should be associated with each alternative. These requirements identify the performance that an alternative is capable of, at a given probability/possibility of exceedance, or risk tolerance. By establishing a risk tolerance for each performance measure independent of the alternative, judgements of performance among the alternatives should be made on a risk-normalized basis. Thanks to such an approach, stakeholders and decision-makers can deliberate the performance differences between alternatives at common levels of risk, instead of having to choose between complex combinations of performance and risk. The NASA RIDM process has been usually simplified as a linear sequence of steps. Essentially, in practice, some steps of the processes may be conducted in parallel, and others may be iterated multiple times before moving to subsequent steps. Particularly, risk analysis of decision alternatives (Part II) is internally iterative as analyses are refined to meet decision needs. Moreover, Part II is iterative with Part III, as stakeholders and decision-makers iterate with the risk analysts to produce a reliable and solid base for taking a robust decision. A simplified algorithm of the decision-making process based on the RIDM concept is shown in Figure 15.2.

The RIDM process is applied to making decisions in safety-critical systems and was adopted by two American regulatory bodies in the complex fields of aerospace and nuclear engineering, namely the National Aeronautics and Space Administration (NASA) and the United States Nuclear Regulatory Commission (US NRC). Even though both regulatory bodies employ the same approach, their interpretation of RIDM seems to be different. The NRC implementation in the nuclear area is more

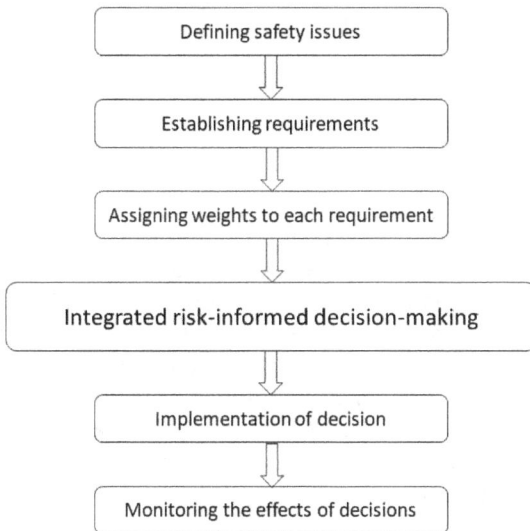

**Figure 15.2:** A simplified algorithm of the decision-making process based on the RIDM concept.

focused on increasing the impact that Probabilistic Risk Assessment (PRA) have on regulatory decisions (e.g., on how to resolve possible conflicts between deterministic analyses and PRA); whereas, the NASA focus on a combination of a better decision structure crossing internal organizational boundaries and support for multi-criteria decisions under uncertainty. Despite this to some extent different interpretation, it can be concluded that in general the main strengths of the use of a probabilistic approach to inform decisions related to safety-critical systems are the following (IAEA, 2005):

- The process of analysis starts from a comprehensive list of initiating events and sets out to identify all the failure sequences that could lead to system damage.
- The frequencies of any initiating event and system or component failure probabilities are included explicitly and not approximately in the PRA model, which makes it possible to determine whether the design is balanced.
- The analysis provides a quantitative estimation of the risk level from the system, and the PRA models: all initiating events, hazards, and structures in a single model, what makes it possible:
  - to derive the relative importance of each of them explicitly. Modern PRA software provides calculations of several important functions that can be used to determine the risk significance of all the initiating events, fault sequences and systems as well as critical components included in the PRA model.
  - to identify also where improvements to the design and operation of the system are needed to provide the greatest reduction in risk.
  - To carry out the process of comparing relative risk values.
- Contemporary PRA software allows some of the parameter uncertainties to be addressed explicitly.

Despite the advantages discussed above, the implementation of the RIDM concept encounters significant difficulties in many cases. A main obstacle is the fact that quantitative risk acceptance criteria for all categories of regulatory objectives do not exist in the most areas of production and services. The top-level safety goals, which give answers to the essential question 'How safe is safe enough?', must be defined by the policy and society. If such a definition is not provided, the elaboration of detailed quantitative acceptance criteria on the safety function, system and component levels by technical expertise is meaningless (Hahn, 2002). In addition, it is very important to ensure that the results of a PRA analysis are not used outside their range of validity. This shortcoming relates to the use of a particular PRA for a particular application and needs to be determined by the user after careful analysis in each situation. In practice, it is also difficult to demonstrate that the PRA model is complete, it means that all relevant initiating events and fault sequences have been identified. Specific problems are posed by situations involving a high level of uncertainty, and thus the need to estimate risks based on subjective expert opinion. Moreover, potentially risk-relevant factors such as the influence of organization, management, and safety culture are usually not considered in this process.

Summarizing the advantages and disadvantages of the RIDM approach, it should be stated that, in risk-informed decision-making, the PRA provides only one of the inputs into the decision-making process. The others' inputs, being related to factors such as the degree to which any mandatory requirements are met, are generally determined using deterministic analysis. Currently a risk-based approach in which the input from the PRA is the one and only input into the decision-making process is not advisable.

## 15.3 Dependability-related Concept of Decision-making

As demonstrated in Section 15.2, effective application of the 'risk-informed decision-making' concept requires, first and foremost, the availability of data on the probabilities of events and scenarios that pose a threat to a given system. For such complex structures as Cyber-Physical-Social Systems, this is mostly impossible, so decision-making about these systems requires a different, more comprehensive approach. It is suggested that this approach should be related to the idea of Dependability Engineering (see Section 10.2). A simplified diagram of the decision-making process based on the dependability-related concept is shown in Figure 15.3.

The method is implemented in 10 steps. At the beginning, the system under investigation should be precisely defined. This means separating it from its environment, defining its boundaries (often vague and fuzzy) and identifying its structure, and then recognizing all relevant processes carried out within this system. Step two boils down to the acquisition and collection of data and information on the system under study, and on this basis build a trustworthy and open knowledge base that could be modified during the research process. Then, a team of experts select *potential sources of risk* (e.g., threats and hazards) based on the literature, data, and their own experience. The set of identified risk sources is the basis for generating potential *risky scenarios*, which (although sometimes with a very low probability) may occur within the analysed system. Potential sources of risk and possible adverse event scenarios are the basis for designing appropriate *security barriers* and firewalls, the role of which is to protect the investigated system from these threats to a minimum.

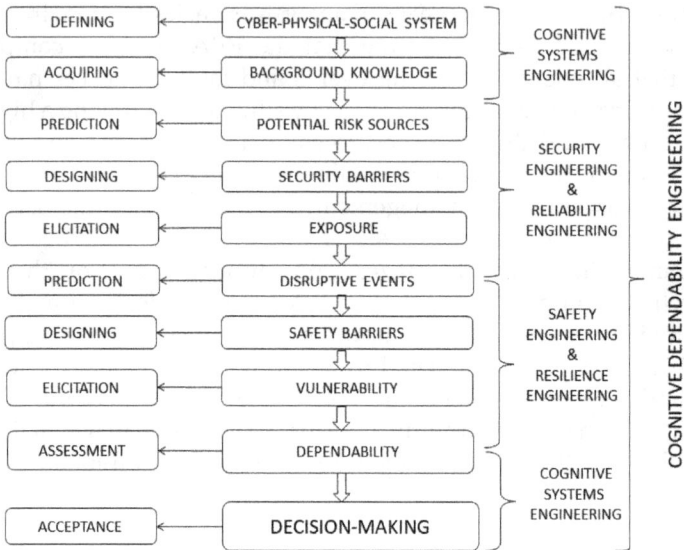

**Figure 15.3:** A simplified diagram of the decision-making process based on the dependability-related concept.

A properly designed Cyber-Physical-Social Systems should be equipped with a complex security system consisting of several proactive barriers, whose role is to minimize the exposure of the system under study to threats and hazards. Thus, the next step is to assess the efficacy of these barriers for each of the potentially possible risky scenarios. These risks, which are not effectively blocked by security barriers, become direct system *exposures*, and are defined as the initiating events that can interrupt the continuity of a process and result in *disruptive events*. Exposure assessment and prediction of the chance of disruptive events can be carried out by *elicitation* by experienced experts.

The next stage of the process is a comprehensive evaluation of resistance, robustness, and resilience of the system to the exposure, the estimation of the susceptibility of the analysed system to the disruptive event and the prediction of the consequences of events in the form of losses caused by these events. The negative effects of disruptive events should be minimized using reactive *safety barriers*. Thus, the next step is to assess the efficacy of these barriers for each of the potentially possible disruptive event, and the calculation of the system vulnerability to each disruptive event. The final stage of the proposed procedure is to assess the overall *risk matrix*, taking into consideration all possible scenarios because of the disruptive events. And finally, *dependability measure* of the investigated system can be expressed by the formula (10.3), described in more detail in Section 10.2.

The concept presented above forms the basis of a generalized cognitive dependability model and Cognitive Dependability Governance Framework for Cyber-Physical-Social Systems operating under deep uncertainty, which will be presented in Chapter 16.

## 15.4 References

Adams, J. (1995). *Risk*. London: UCL Press.

Aven, T. (2012). The risk concept: Historical and recent development trends. *Reliability Engineering and System Safety*, 99: 33–44.

Aven, T. (2020). *The Science of Risk Analysis. Foundation and Practice*. London and New York: Routledge, Taylor, and Francis Group.

Bukowski, L. (2019). *Reliable, Secure and Resilient Logistics Networks. Delivering Products in a Risky Environment*. Springer Nature Switzerland AG 2019. ISBN 978-3-030-00849-9 (Hardcover), ISBN 978-3-030-00850-5 (eBook).

de Moivre, A. (1738). *Doctrine of Chances*. New York: Chelsea Publishing.

Dezfuli, H., Stamatelatos, M., Maggio, G. and Everett, C. (2010). Risk-informed decision-making in the context of NASA risk management. *In: Proceedings of the PSAM 10 Conference*, Seattle, WA.

Hardy, C.O. (1923). *Risk and Risk Bearing*. Chicago: University of Chicago.

IAEA. (2005). *Risk-informed Regulation of Nuclear Facilities: Overview of the Current Status* (IAEATECDOC-1436). Technical report, IAEA, Vienna. Available at http://www.pub.iaea.org/MTCD/Publications/PDF/TE_1436_web.pdf.

ISO. (2009). *Risk Management – Principles and Guidelines*. ISO 31000: 2009.

Kaplan, S. and Garrick, B.J. (1981). On the quantitative definition of risk. *Risk Analysis*, 1: 11–27.

Knight, F.H. (1921). *Risk, Uncertainty, and Profit*. Boston: Houghton Mifflin Co.

NASA. (2008). *Agency Risk Management Procedural Requirements* (NPR 8000.4A). Technical report, NASA, Washington, DC. Available at http://nodis3.gsfc.NASA.gov/displayDir.cfm? Internal_ID=N_PR_8000_004A_47.

NASA. (2010). *Risk-informed Decision making Handbook* (NASA/SP-2010-576). Technical report, NASA. http://standards.NASA.gov/documents/viewdoc/3315763/ 3315763.

Rumsfeld, D. (2011). *Known and Unknown: A Memoir*. New York: Penguin Group.

Sornette, D. (2009). Dragon-kings, Black swans, and the prediction of crises. *Int. Journal of Terraspace Science and Engineering*. http://www.arxiv.org›physics.

SRA Glossary. (2015). http://www.sra.org/sites/default/files/pdf/SRA-glossary-approved 22 june2015-x.pdf.

Taleb, N.N. (2010). *The Black Swan: The Impact of the Highly Improbable*. London: Penguin Books.

Zio, E. and Pedroni, N. (2012). *Risk-informed Decision-making Processes. An overview*. https://www.researchgate.net/publication/266391225_Overview_of_risk-informed_decision-making_processes.

# 16

# Cognitive Dependability Based Problem-solving in Cyber-Physical-Social Systems Operating Under Deep Uncertainty*

Decision-making under conditions of uncertainty is an important and responsible activity, being the final step in the problem-solving process. In real-world practice, the entire process is fraught with uncertainty, with the degree of uncertainty having a significant impact on the risk involved in making the final decision. In the first part of the chapter, general concepts and principles related to the consideration of uncertainty in the problem-solving process are presented with particular emphasis on the decision-making stage. The second part is devoted to the same issues related to conditions of deep uncertainty. It has been proposed to use the concept of a digital twin to support the process of problem solving under deep uncertainty using an interactive dialogue system between a human expert, representing natural intelligence, and a cognitive digital twin, representing artificial intelligence. The last section of the chapter presents a framework for cognitive dependability based problem-solving in Cyber-Physical-Social Systems operating under deep uncertainty. The starting point for the development of this framework is a simplified model of a Cyber-Physical-Social System cognitive dependability, and a general

* https://orcid.org/0000-0002-2630-3507.

model for solving operational problems under deep uncertainty based on the concept of cognitive dependability, which are presented in both graphic and descriptive form.

## 16.1 Decision-making Under Uncertainty

*Decision-making* is the cognitive process of choosing between two or more alternatives, ranging from the relatively clear cut to the complex (see Section 5.2). It is the last link in a chain composed of basic, meta, and higher cognitive functions. All these cognitive functions, along with previously accumulated background knowledge, form the basis for making a certain decision in each situation. Since in practice our background knowledge is incomplete and cognitive functions are imperfect, therefore, it can be assumed that the decision-making process is carried out under conditions of uncertainty. The concept of uncertainty is understood as a situation such as:

- the order, nature or state of things is unknown, and/or
- the consequence, extent, or magnitude of circumstances, conditions, or events is unpredictable.

The degree of this uncertainty can vary and is traditionally assigned to two groups, namely, aleatory and epistemic uncertainty (e.g., Smith, 2014). The aleatory uncertainty is understood as an inherent variation associated with the given system or its environment. It can be observed in random experiments and described by probability distributions. Traditional reliability engineering and risk analysis applications tend to model only the aleatory uncertainties, which can lead to significant underestimations of the real risks and overestimation of reliability. However, the epistemic uncertainty is not an inherent property of the system or its environment, and it results from our inability to understand as well as describe and model the reality. Thus, in this case, the standard probabilistic methods are not useful.

A different approach to the division of uncertainty is presented in their work by Kay and King (2020), proposing a distinction between resolvable and radical uncertainty. *Resolvable uncertainty* can be removed by additional observations or tests and described by a specific probability distribution. In contrast, *radical uncertainty* is more complex and can have multiple dimensions, such as obscurity; ignorance; vagueness; ambiguity; ill-defined problems; and a lack of information (Kay and King, 2020). This problem is discussed in more detail in Section 5.3 of this book. An extreme case of radical uncertainty is *deep uncertainty*, which is defined in Section 15.1 and as such cannot be described in probabilistic terms.

Classical methods of decision-making under uncertainty are based on two basic assumptions, namely that uncertainty belongs to the 'aleatory' category, and therefore has a random character described by a certain probability distribution, and that the decision-maker is guided by principles of rationality. In the case of deep uncertainty, neither of these conditions is met, especially the assumptions formulated in von Neumann and Morgenstern's (1953) theory. They assumed that the preferences over alternative probability distributions are (Kay and King, 2020):

- complete, meaning that the decision-maker is able to choose between all possible probability distributions;
- transitive, which means if X prefer A to B and B to C then X prefer A to C;
- continuous, which means if A is preferred to B, and B to C, there is always some situation involving A and C which will be preferred to B, and this is true whether A, B and C are fixed outcomes or probability distributions;
- independent, meaning if A is preferred to B, that preference of A over B is maintained regardless of other options available.

These assumptions can only be met for uniquely defined problems for which event frequencies are given or can be determined with acceptable accuracy. Thus, it does not apply to deep uncertainty type situations, so the classical methods of decision-making under uncertainty should not be used in these circumstances.

An approach more appropriate for describing exceptional situations is the concept of *Value at Risk* (VaR), based on portfolio theory pioneered by Markowitz in 1952 and developed by Morgan in the late 1980s (a detailed description of the method with mathematical foundations and application examples is included in the book (Morgan, 1998)). While it requires knowledge of the parameters of the probability distribution spread (variance and covariance), it provides a means of assessing the frequency of rare or unlikely events in the form of percentiles (e.g., an event occurring 1 time in 1,000 cases). In engineering, the equivalent of this method is the use of extreme value theory to determine, for example, the so-called 'century wave'. The disadvantage of these methods is that they require a very large number of data, i.e., a representative random sample, which is impossible to meet in the case of deep uncertainty.

The question then arises, how to 'tame' the problem of uncertainty in this situation, when we cannot use probabilistic methods? Some guidance, and a starting point for further consideration, may be the division of problem-solving methods into two types proposed by Greg Treventon (2007): puzzles and mysteries. *Puzzle-type problems* are characterized by well-defined rules, an unambiguous solution, and knowledge of how to solve the problem, and are therefore solvable. Unlike puzzles, *mystery-type problems* cannot be clearly defined, so they cannot be objectively solved.

They can only be limited by identifying the critical factors affecting the object to be decided and examining possible interactions between these factors in the future under the influence of possible changes in the environment. On this basis, it is possible to make decisions incrementally, looking for better and better solutions and rejecting relatively worse ones, being aware of the incompleteness and imperfection of the knowledge possessed by the decision-maker.

To summarize, it can be said that deep uncertainty belongs to the 'mystery' type of problems and has an 'unknown-unknown' character. It mainly applies to systems of high complexity, such as Cyber-Physical-Social Systems, which are often characterized by reflexivity, whose environment changes discontinuously and unpredictably, and whose decision-maker acts within the limited rationality, under the influence of emotions and is subject to cognitive illusions. So, none of the above methods meet the conditions for decision-making under deep uncertainty, therefore there is a need to develop a different approach to this problem.

## 16.2 Problem-solving Under Uncertainty

Based on widely recognized dictionaries, it is proposed to adopt the following definition of the term *problem*: a situation, matter, or question involving doubt or uncertainty that is difficult to deal with or understand. Any problem has at least three main components: givens, goal, and operations. *Givens* are the facts or pieces of information presented to describe the problem. *Goal* is the desired end-state of problem solving. *Operations* are the actions to be performed in reaching the desired goal (Newell and Simon, 1972). Problems can be categorized as ill-defined or well-defined, based on how a given problem is represented. Problems with complex representations and/or many solutions are called *ill-defined*. Problems with discrete representations and one solution are named *well-defined*. The distinction between ill-defined and well-defined is blurred and fuzzy, based on the complexity of the problem and what is required cognitively to solve it.

A general definition of 'problem solving' is as follows: the process or act of finding a solution to a problem (https://www.merriam-webster.com/dictionary/problem-solving).

More specifically, it can be described as 'the way by which solutions are developed to remove an obstacle from achieving an ultimate goal'. In engineering, the term 'problem solving' is used when products or processes fail, and corrective action can be taken to prevent further failures (https://en.wikipedia.org/wiki/Problem_solving).

The fundamental difference between decision-making and problem-solving can be boiled down to the following statement: problem solving is the process of investigating the given information and finding all possible solutions through invention or discovery, and it is an important step toward making the right decisions. Thus, the information gathered in that process may be used towards decision-making (https://en.wikipedia. org/wiki/Decision-making).

Typical problems and their causes can be boiled down to the following characteristics:

- Problems are usually deviations from performance standards.
- Problems should be precisely identified and described.
- Problems are caused by changes caused by changes in the environment or system operating conditions.
- It is necessary to distinguish between effects caused by the problems in question and those arising from other causes.
- Potential causes of problems should be inferred from relevant changes found in analysing the given problem.
- It is recommended to choose such a cause of the problem that exactly explains all the facts, while having the fewest (or weakest) assumptions.

Problem-solving knowledge is of two kinds. *Declarative knowledge* is knowing that something is the case, and concerns facts, theories, events, and objects. *Procedural knowledge* is knowing how to do something. It involves such cognitive properties as motor skills, cognitive skills, and cognitive strategies. Both declarative and procedural knowledge are stimulated in working memory as problem solving occurs, as well as the two forms of knowledge are both distinct and interdependent. Declarative and procedural knowledge interact in many ways during problem solving. A basic component of declarative knowledge in the human information-processing system is the proposition. It expresses or proposes the relationships among different concepts (Hardin, 2003).

It is generally assumed that there are three attributes that are commonly used to differentiate expert from novice problem-solving characteristics. These attributes include conceptual understanding; basic, automated skills; and domain-specific strategies. *Conceptual understanding* refers to both the actual information in memory and the organization of that information in memory. Information is to be stored in memory as patterns or structures that, once instantiated, creates a principle for evaluating any new information. Having a conceptual understanding of a domain means that an individual can make meaning of domain-specific situations or problems, based on *prior knowledge* of that domain.

*Basic, automated skills* in any domain allow an individual to perform necessary and routine operations without including the thinking process.

These skills are overlearned to the point that they become habitual and sometimes unconscious, enabling individuals to operate quickly and accurately without demanding their short-term memories. Automaticity allows people to focus their attention on the more complex tasks associated with a specific domain and is a general attribute associated with expert activities in a domain, as well as supports the expert's speed and skill of execution.

Unlike basic, automated skills, *domain-specific strategies* persist under conscious control. They are the processes and procedures in a domain that an expert should consciously think about to solve a problem. Therefore, they are the procedural knowledge associated with a domain. Expert–novice differences have been studied and described within the context of three main attributes: experts exhibit better conceptual understanding of their domain, use more automated skills and domain-specific strategies, and have a conceptual understanding that is declarative, while basic skills and strategies are procedural (Hardin, 2003).

Rational problem solving requires the use of effective methods and techniques. A particularly popular technique, known as *general problem-solving strategy*, is the method proposed by George Polya (2014), which is based on four principles that represent successive steps in the problem-solving process. The following is an abbreviated description of the steps in this process:

1. Understand the problem

To help understand a complex problem, Polya recommends breaking it down into simpler components using the following questions:

What is the unknown? What are the data? What is the condition? Is it possible to satisfy the condition? Is the condition sufficient to determine the unknown? Or is it insufficient? Or redundant? Or contradictory?

The results of the answers to these questions should be presented as follows:

Draw a figure. Introduce suitable notation. Separate the various parts of the condition. Write them down.

2. Devising a plan

The starting point for this stage is to look for the relationship between the information we have and what is unknown. It is suggested to use the following questions as a signpost to find this relationship:

Have you seen it before (or have you seen the same problem in a slightly different form)? Do you know a related problem or a theorem that could be useful? Could you use it, or could you use its result? Could you use its method? Should you introduce some auxiliary element to make its use possible? Could you restate the problem?

If you cannot solve the proposed problem, try to first solve some related problem to answer the following questions:

Could you imagine a more accessible related problem? A more general problem? A more special problem? An analogous problem? Could you solve a part of the problem? Could you derive something useful from the data and information? Could you think of other information appropriate to determine the unknown? Could you change the unknown and information, so that the new unknown and the new information are nearer to each other? Did you use all the information? Did you use the whole condition? Have you considered all essential notions involved in the problem?

### 3. Carrying out the plan

Carrying out your plan of the solution, check each step. Check the following: Can you see clearly that the step is correct? Can you prove that it is correct?

### 4. Looking back

Examine the solution obtained. Try to answer the following questions: Can you check the result? Can you check the argument? Can you derive the solution differently? Can you see it at first sight? Can you use the result, or the method, for some other problem?

General problem-solving strategies have also been called *heuristics* and are the key to problem-solving expertise and intellectual performance (a frequently used synonym for heuristics is rule of thumb). The heuristic methods can be applied to a problem in any area hence they are general problem-solving skills. People often must make decisions in the face of uncertainty, with vague information about the problem, based on suggestive but inconclusive evidence. The reasoning processes used to resolve the uncertainty are named *judgement heuristics*. One form of judgement heuristic is similarity judgement, where an example is evaluated based on prior knowledge of a similar case. A comparable type of judgement is *representativeness*, where an assumption is made that the characteristics of the individual cases are representative of entire group. Another heuristic is the *availability heuristic*, where evaluations are made on the elements most readily available in memory. *Analogical reasoning* is another heuristic method, where judgement is made by search for similarities to events that have occurred previously. Heuristics also include the creation of *mental models*, e.g., to predict the outcome of an event. All these heuristics are examples of general-purpose thinking skills, which can be applicable to many practical problems. The heuristics approach emphasizes finding a good representation of the problem. While content-specific knowledge is required to solve the problem, the general problem-solving skills are also valuable by studies in the domains of mathematics and computer science (Hardin, 2003).

The idea that problem-solving success was directly related to general problem-solving skill was supported by some works in artificial intelligence. A theory that emphasized the similarities between artificial intelligence and human problem solving is based on four underlying principles (Hardin, 2003):

- a small number of main characteristics of the problem-solving process are invariant over the task and the problem solver,
- the characteristics of the problem are sufficient to determine the entire problem space,
- the structure of the task environment determines the possible structure of the problem space,
- the structure of the problem space determines the possible methods that can be used for problem solving.

However, in many cases general problem-solving knowledge was determined to be an incomplete explanation of how problem solving occurred. Several studies disclose that expertise relies on both domain-specific knowledge and problem-solving skill. For example, Bruer (2016) states that expertise relies on highly organized, domain-specific knowledge that can arise only after extensive experience and practice in the domain. Studies in domain-specific problem-solving expertise moreover present the underlying principle of metacognition. *Metacognition* is the ability 'to think about thinking', the self-awareness of problem solving, and the ability to monitor and control one's mental processing. The ability to solve problems successfully depends on many factors related to the human information-processing system. Summarizing the years of research in this area it can be said that six main characteristics of expert performance are of particular importance (Chi et al., 1988):

- Experts perceive large, meaningful patterns in their domain. This ability to see meaningful patterns reflects organization of the knowledge base.

- Experts are faster and more accurate than novices at solving problems within their domain because they have developed the basic, automated skills applicable to the problem, and they have an organized database from which to retrieve the solution.

- Experts have superior short- and long-term memory, based on superior memory organization, rather than large volume.

- Experts can see and represent data at a more conceptual level than novices.

- Experts usually spend more time analysing and evaluating a problem quantitatively before beginning to solve the problem.
- Experts have strong self-monitoring skills. They are more aware when they make errors or mistakes, why they fail to comprehend, and when they need to check their solutions.

From the above, it follows that both content knowledge and general problem-solving skill are necessary for expert problem solving to occur. Content-specific knowledge allows the expert to perceive information in a way that maximizes memory and information is conceptualized more on the level of principles, whereas, superior problem analysis, assessment and strong self-monitoring skills can be recognized as general problem-solving expertise.

## 16.3  Complex Problem-solving Under Deep Uncertainty – The Concept of Cognitive Digital Twin

The general strategy-based problem-solving methods described in Section 16.2 work well in practice if the problems are not too complicated, and the uncertainty is moderate. For problems of high complexity considered under deep uncertainty, a different approach is necessary, the concept of which will be described below. It is proposed to use the idea of a *digital twin* to create a problem-solving support system for cognitive processes leading to solutions that fully satisfy the decision-maker.

In recent years, most organizations have been exposed to more fast-paced, uncertain, and complex boundary conditions, as well as the pressure to achieve ever higher productivity and efficiency in the shortest possible time. To meet these challenges, the idea of so-called Digital Twin has developed over the last two decades. Generally, this term describes the virtual representation of a real system with possibility to establish a bidirectional coupling between the digital and the physical system (Grieves and Vickers, 2017). Eckert et al. (2019) identified complete integrated twins as one of the major industry trends to 2040. The main benefits of the concept lie in an increased accuracy and fidelity as well as decreased time, costs, and work expenditure in the process of creating real-time virtual representation of real systems (Jones et al., 2019). Digital Twins were primarily applied in production or after-production stages, rarely in earlier stages such as product design (Tao et al., 2019).

In 2010, the term Digital Twin appeared in the definition of the NASA as an 'integrated multi-physics, multiscale, probabilistic simulation of a vehicle or system that uses the best available physical models, sensor updates, fleet history, etc., to mirror the life of its corresponding flying twin' (Shafto et al., 2010). In contrast, in later years the term was adapted to the needs of the manufacturing industry as: "A digital representation of an active unique product ... or unique product-service system ... that comprises its selected characteristics, properties, conditions, and behaviors by means of models, information, and data within a single or even across multiple life cycle phases" (Stark and Damerau, 2019). Schleich et al. (2017) suggest using an abstract conceptual model and a virtual representation, which makes the model scalable, interoperable, expansible, and enables it to achieve high fidelity. Digital Twin is often described as a higher level of simulation, including real use phase data (Boschert and Rosen, 2016; Schluse and Rossmann, 2016). A Digital Twin can exist over the complete lifecycle, subdivided in the phases 'as-designed', 'as-built', and 'as-maintained' (Eigner et al., 2019).

There are also other terms, which are to some extent used as synonyms, such as 'Digital Blueprint' (e.g., Bajaj et al., 2016), 'Digital Mirror Model' (e.g., Erikstad, 2017), 'Digital Thread' (e.g., Siedlak et al., 2018) and 'Device or Digital Shadow' (e.g., Schuh and Blum, 2016; Baltes and Freyth, 2017). Whereas the 'Twin' focusses on phases of operation and service, the 'Thread' addresses earlier phases of acquisition and design, the term Digital Shadow is often used synonymously, but it cannot give feedback to the real system, so a bidirectional data exchange would be not possible.

Based on the before described meanings from the literature review, there are three essential characteristics of Digital Twins, which should be included in the definition (Trauer et al., 2020):

- The Digital Twin is a virtual dynamic representation of a physical artefact or system,
- Data is automatically and bidirectionally exchanged between the Digital Twin and the physical system,
- The Twin entails data of all phases of the entire product lifecycle and is connected to all of them.

Accordingly, the following definition was proposed by Trauer et al. (2020): "A Digital Twin is a virtual dynamic representation of a physical system, which is connected to it over the entire lifecycle for bidirectional data exchange."

It is proposed to use the concept of a digital twin to create a system for solving complex problems under deep uncertainty according to the model shown in Figure 16.1. The proposed concept also considers the results of research in neuro-psychological sciences and artificial intelligence (e.g., Adams and Aizawa, 2017; Rescorla, 2020).

This system consists of two parallel, interactive subsystems that communicate with each other at all stages of the problem-solving process.

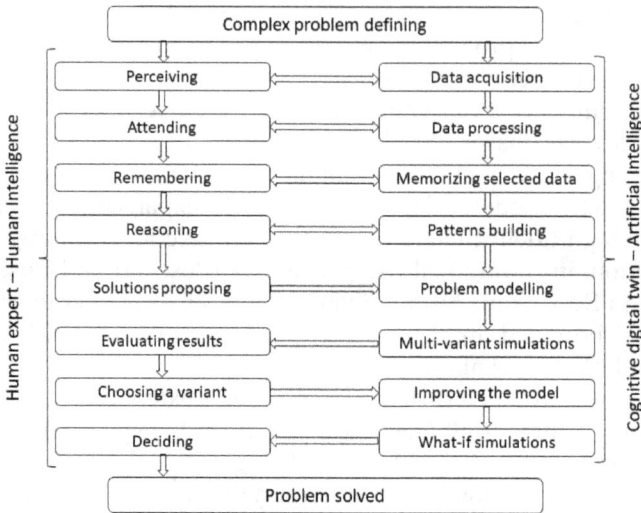

**Figure 16.1:** A system for solving complex problems under deep uncertainty – a conceptual model.

One of them represents human intelligence and is carried out by an expert or expert group. The other subsystem is implemented based on artificial intelligence and has been named as a cognitive digital twin. The process begins with defining the problem to be solved in such a way that it is clearly understood by both human experts and the digital subsystem. The next four steps in the process run in parallel in both subsystems, with each step involving an exchange of information between the subsystems.

The first step of the human expert's problem-solving process is 'perceiving', understood more broadly than 'perception' in everyday language. It includes, in addition to the typical capabilities of visual activities, such as: sensory integration within any one sensory modality achieved by connections within that system; integration between sensory modalities achieved by connections to multimodal areas; higher-order areas form representations that are invariant by integrating information from lower-order areas; as well as 'salience networks' – areas responding when a salient event is detected (Passingham and Rowe, 2016). The counterpart of this step in the digital area is data acquisition, also known as collection, gathering, or mining. The mutual exchange of information illustrated by the two-way arrow allows for iterative enrichment of the available data by both the expert and his digital twin. The mutual exchange of information illustrated by the bidirectional arrow allows for iterative enrichment of the available data by both the expert and his digital twin, which proceeds in a similar manner in the next three stages as well. On the expert side are typical cognitive functions, and on the digital side are their counterparts in processing and using data to acquire new knowledge. The result of these steps should be that the expert generates several (and sometimes more than a dozen) proposals for solving the problem, which are the starting point for building models describing these solutions. The cognitive digital twin in the next step performs numerous multivariate simulations on these models and directs their results to the human expert for evaluation. The expert selects the most favourable solution variant and directs it to the twin for modification and improvement of the model, on which it performs multiple 'what-if' simulations. The results of these simulations are evaluated by the expert and a final decision is made on how to solve the problem. In the case of a positive decision, the problem can be considered solved.

## 16.4 A Framework for Cognitive Dependability Based Problem-solving in Cyber-Physical-Social Systems Operating Under Deep Uncertainty

Taking the concept of Cognitive Systems Engineering (see Section 11.3) as a starting point as well as the dependability-related concept of decision-making (see Section 15.3), a simplified model of Cyber-Physical-Social Systems cognitive dependability is proposed (shown in Figure 16.2).

In the structure of this complex system, three basic systems are distinguished, namely: a physical system composed of *n* subsystems, whose main operating characteristic is reliability; a cybernetic system characterized by three basic properties – confidentiality, integrity, and

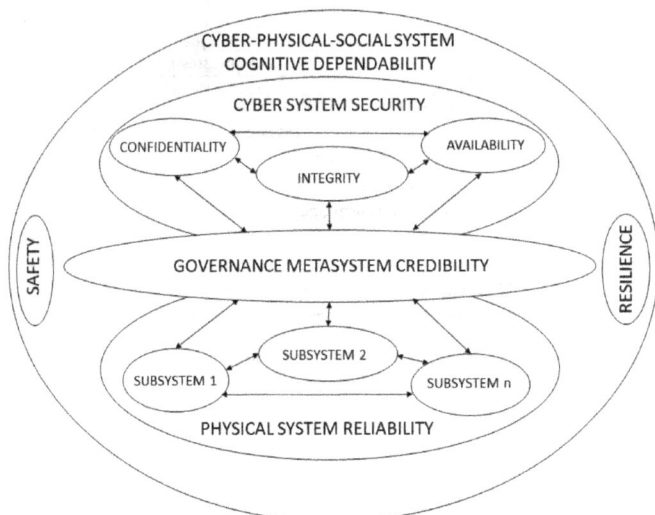

**Figure 16.2:** A simplified model of a Cyber-Physical-Social System cognitive dependability.

availability only for authorized persons, and governance metasystem described by its trustworthiness. Two further operational features emerge from the interaction between these systems during the operation of the entire system, namely safety and resilience. According to the assumptions made earlier, these five properties can be reduced to one general concept – cognitive dependability, which is an amalgam of all relevant operational characteristics of complex Cyber-Physical-Social systems.

Using the cognitive digital twin concept described in the previous section and the proposed system for solving complex problems under deep uncertainty, a generalized model was developed in the form of a framework for solving operational problems under deep uncertainty based on the concept of cognitive dependability (Figure 16.3).

The process of solving complex operational problems under deep uncertainty consists of seven steps. The first is to formulate all the relevant conditions and constraints for the problem, with particular attention to potentially possible (including those that have not previously existed) risks. These should be defined both in the form of random events of the disturbance type and hazard scenarios that can lead to cascade-type phenomena. For particularly complex problems and high level of uncertainty, both at this stage and the next, the algorithm presented in Section 16.3 and illustrated in Figure 16.1 can be used.

The second stage involves developing a reliability assessment as well as security for a system composed of physical and cyber parts.

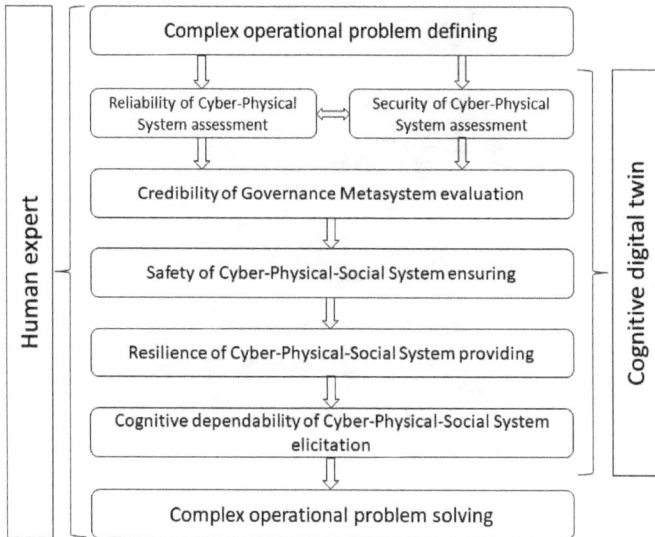

**Figure 16.3:** A general model for solving operational problems under deep uncertainty based on the concept of cognitive dependability.

Since reliability refers to normal conditions (i.e., those predicted to be acceptable in operation), and security has a significant impact on under which conditions the object under study will be used, therefore the two parameters are closely related. Any change in security barriers can significantly affect the reliability of the system, so stage two should proceed in an iterative manner and be completed only after satisfactory solutions have been obtained on both the reliability and security sides. The basic rules, principles, methods, and models used in reliability assessment are included in Chapter 6 (Reliability Engineering – the concept of failure-free operation), while security is discussed in Chapter 8 (Security Engineering: The Concept of Cyber-security).

Stage three concerns the evaluation of the social part of the entire system, specifically the metasystem responsible for the proper governance of the Cyber-Physical-Social System. This is a particularly important element of the whole, and at the same time difficult to evaluate objectively, so it is proposed to adopt the concept of 'credibility' as a measure of the quality and confidence in the operation of this system. So far, there is a lack of quantitative measures of this property, therefore it is proposed to use linguistic variables of the type: very low, low, medium, high, and very high, to evaluate it. The problem of providing an adequate level of credibility to the governance metasystem is poorly represented in the subject literature. The paper (Andrzejczak and Bukowski, 2021) reviews methods for assessing the quality and credibility of expert evaluations and proposes an author's procedure for verifying the accuracy of expert evaluations a using probability theory and mathematical statistics.

The next two stages are concerned with shaping the emergent features of the entire Cyber-Physical-Social System, a highly challenging process due to the circular feedbacks that occur in highly complex systems. The result of this feedback is, among other things, the difficulty of determining cause-effect chains, because in many cases it is not possible to determine what is a cause in it while what is an effect. The implementation of these steps must therefore proceed gradually, in many steps, based on the principles of Safety Engineering and Resilience Engineering (see Chapters 7 and 9) in an iterative manner supported by a dialogue between a human expert and a cognitive digital twin. However, in the case of high risk and deep uncertainty, classical methods guarantee neither adequate safety nor system resilience. Therefore, in these situations, it is proposed to use the precautionary principle to ensure a satisfactory level of security and an extended resiliency matrix to ensure the required level of system operation continuity.

The *precautionary principle* can be put in simple terms as follows: "When an activity raises threats of harm to human health or the environment, precautionary measures should be taken even if some cause-and-effect relationships are not fully established scientifically." This principle provides the foundation for European environmental law and plays an increasing role in developing environmental health policies as well (Hayes, 2005). The European Commission states that:

Recourse to the precautionary principle presupposes that potentially dangerous effects deriving from a phenomenon, product or process have been identified, and that scientific evaluation does not allow the risk to be determined with sufficient certainty. The implementation of an approach based on the precautionary principle should start with a scientific evaluation, as complete as possible, and where possible, identifying at each stage the degree of scientific uncertainty (EC, 2000).

The European Commission further specifies the requirements to which this approach should conform, namely:

- start with the fullest possible scientific evaluation, identifying at each stage and as far as possible the degree of scientific uncertainty;
- entail an evaluation of various risk management options, including the option of taking no precautionary action;
- involve as early as possible and, to the extent reasonably possible, all interested parties.

Any measures adopted based on the precautionary principle should also be proportionate, non-discriminatory, and consistent, based on cost-benefit analysis, and subject to review when new credible and valuable information becomes available. So, the main advantage of the approach is that it provides guidance on a process for avoiding serious or irreversible harm under conditions of deep uncertainty and has wide recognition in international Law. The main shortcoming of the precautionary approach is its possible contradiction with the principle of value creation based on CBA (Cost-Benefit Analysis). In such situations, special attention should be paid to the proper balance between the goals of achieving a high level of safety and profitability, by seeking a reasonable compromise between both (Aven, 2020).

The fifth stage is a consequence of the previous stages, and its task is to ensure the continuity of the entire system if the preventive actions of the previous stages are ineffective. For this purpose, resilience engineering tools will be used, based on the concept of extended resilience matrix. This concept is based on assigning appropriate activities and actions to the various phases of the system process. The assumptions of this concept are based on a typical time course of system operation disturbed by a disruptive event, shown in Figure 16.4.

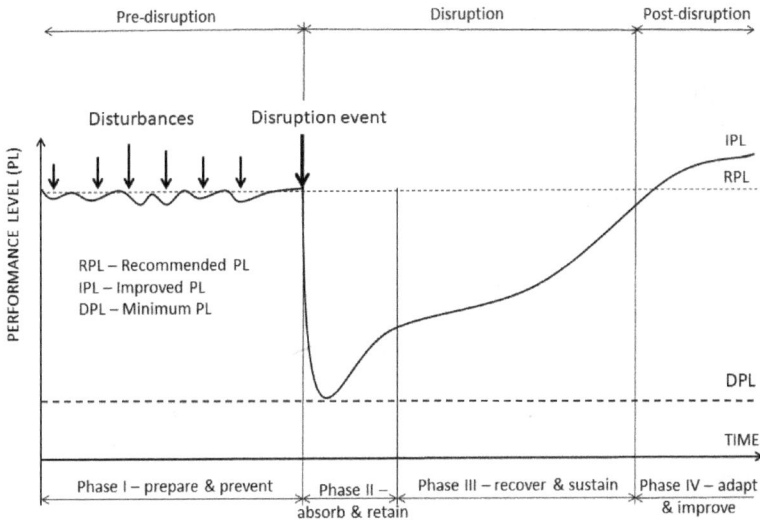

**Figure 16.4:** A typical time course of system operation disturbed by a disruptive event.

Three basic periods can be distinguished in this course, namely, Pre-disruption, Disruption, and Post-disruption, which in turn are divided into four phases:

- Phase I – preparing and prevent based on hindsight and foresight analysis (HFA);
- Phase II – absorb and retain (responding to disruption event by using buffering capacity and survivability);
- Phase III – recover and sustain (restoring functionality by using restructuring ability and redundancy), and
- Phase IV – adapt and improve (increasing performance by using new knowledge and learnability).

Ten activities were singled out that should support the achievement of the set goals, namely:

1. Planning – system description by definition and specification:
   a. Critical Functions (CF) and their parameters,
   b. Threshold Values (TV) for these parameters.
2. Identifying – recognition of potential risk sources and scenarios leading to disruption.
3. Protecting – preservation against threats and hazards by safety and security barriers.
4. Monitoring – supervision of all signals that significantly affect critical system functions.

5. Detecting – discovering deviations in signals indicating significant changes in the system.
6. Responding – reducing (mitigation) the consequences of damage.
7. Repairing – removal of damage and restoration.
8. Learning – creating new knowledge about the system.
9. Evolving – modification of the system.
10. Anticipating – short- and long-term forecasting.

Combining the above-mentioned phases of the exploitation processes with the corresponding activities, one gets the so-called extended residual matrix, which is shown in Table 16.1.

The detailed design of individual activities for a specific practical case should be preceded by a "what-if" analysis conducted in dialogues between a human expert and his digital twin.

The final stage of the cooperation of the human expert with his virtual partner, the cognitive digital twin, is the sixth step, the aim of which is to elicit the level of dependability of the entire analysed complex system. Based on the results of the assessments of the previous stages, the expert assisted by his 'twin' elicits a cumulative measure of the quality and operational dependability of the assessed system in all possible, including unpredictable, working conditions. The basic rules, principles, methods, and models used in area of cognitive dependability is included in Chapter 6 (Cognitive Dependability Engineering – The Concept of Trustworthy Performance). Since for multidimensional attributes such as dependability is generally subjective, so a different approach is suggested for cognitive dependability assessment of Cyber-Physical-Social Systems

**Table 16.1:** Extended resilience matrix: An example.

| Phase → Activity ↓ | Pre-disruption PP – Prepare & Prevent | Disruption AR – Absorb & Retain | Disruption RS – Recover & Sustain | Post-disruption AI – Adapt & Improve |
|---|---|---|---|---|
| 1. Planning | X | | | |
| 2. Identifying | X | | | |
| 3. Protecting | X | | | |
| 4. Monitoring | X | X | X | X |
| 5. Detecting | X | X | X | X |
| 6. Responding | X | X | X | |
| 7. Repairing | X | | X | |
| 8. Learning | X | X | X | X |
| 9. Evolving | X | | | X |
| 10. Anticipating | X | X | X | X |

operating under deep uncertainty. By analogy with the ALARP (As Low as Reasonably Practicable) principle commonly used in risk management, it is proposed to adopt in Cognitive Reliability the AHARP (As High as Reasonably Practicable) principle, which is its opposite (antonym). This means that it is necessary to define and establish according to activity 1 in step five Critical Functions (CF) and their parameters, and Threshold Values (TV) for these parameters for the operational process being evaluated. On this basis, the results obtained in steps two through five should be evaluated according to the criterion of whether the following conditions are met:

- As Reliable as Reasonably Practicable, and
- As Secure as Reasonably Practicable, and
- As Credible as Reasonably Practicable, and
- As Safe as Reasonably Practicable, and
- As Resilient as Reasonably Practicable.

It is theoretically possible to quantify the degree to which the above requirements are met, but in situations involving deep uncertainty, this does not seem reasonable. This is due to the practical principle that "it is better to be inaccurately right than to be exactly wrong".

The seventh stage involves summarizing all the previous stages and making final decisions on how to solve the operational problem. If a solution is found that satisfies the 'owner' of the problem at this stage, the process is completed. If, on the other hand, no positive decision can be made, the whole process should be repeated, with particular attention to the first stage, i.e., redefining the problem. Attempts to demonstrate the usefulness of the proposed method are presented in Chapter 17 using the example of three disasters that have occurred in recent decades and have been analysed in detail by committees of eminent experts.

## 16.5 References

Adams, F. and Aizawa, K. (2017). Causal theories of mental content. *In*: Zalta, E. (ed.). *The Stanford Encyclopedia of Philosophy* (Summer 2017 Edn.). https://plato.stanford.edu/ archives/ sum2017/entries/content-causal/.

Andrzejczak, K. and Bukowski, L. (2021). A method for estimating the probability distribution of the lifetime for new technical equipment based on expert judgement. *Eksploatacja i Niezawodność – Maintenance and Reliability*, 23(4): 757–769. http://doi.org/ 10.17531/ ein.2021.4.18.

Aven, T. (2020). *The Science of Risk Analysis. Foundation and Practice*. London and New York: Routledge, Taylor, and Francis Group.

Bajaj, M., Cole, B. and Zwemer, D. (2016). Architecture to Geometry: Integrating system models with mechanical design. *SPACE Conferences and Exposition: AIAA SPACE 2016*, Long Beach, California. https://doi.org/10.2514/6.2016-5470.

Baltes, G. and Freyth, A. (2017). Die radikal neuen Anforderungen unserer Zeit und die Konsequenz für Veränderungsarbeit. pp. 1–79. *In*: Baltes, G. and Freyth, A. (eds.).

*Veränderungsintelligenz: Agiler, innovativer, unternehmerischer den Wandel unserer Zeit meistern* Guido Baltes, Antje Freyth (Hrsg.), Vol. 131, Wiesbaden, Germany: Springer Gabler. https://doi.org/10.1007/978-3-658-04889-1_1.

Boschert, S., Heinrich, C. and Rosen, R. (2018). Next generation digital twin. Paper presented at *TMCE 2018*, 7– 11 May 2018, Las Palmas de Gran Canaria, Spain.

Bruer, J.T. (2016). Where is educational neuroscience? *Educational Neuroscience*, 1: 1–12.

Bukowski, L. (2019). *Reliable, Secure and Resilient Logistics Networks. Delivering Products in a Risky Environment*, Springer Nature Switzerland AG 2019. ISBN 978-3-030-00849-9 (Hardcover). ISBN 978-3-030-00850-5 (eBook).

Chi, M.H., Glaser, R. and Farr, M.J. (1988). *The Nature of Expertise*. Hillsdale, NJ: Lawrence Erlbaum.

EC. (2017). The precautionary principle: Decision-making under uncertainty. September 2017 Issue 18. https://ec.europa.eu/environment/integration/research/newsalert/pdf/precautionary _principle_decision_making_under_uncertainty_FB18_en.pdf.

Eckert, C., Isaksson, O., Hallstedt, S., Malmqvist, J., Öhrwall Rönnbäck, A. and Panarotto, M. (2019). Industry Trends to 2040. *Proceedings of the Design Society: International Conference on Engineering Design*, 1(1): 2121–2128. https://doi.org/10.1017/dsi.2019.218.

Eigner, M. et al. (2019). Definition des Digital Twin im Produktlebenszyklus. *ZWF Zeitschrift für wirtschaftlichen Fabrikbetrieb*, 114(6): 345–350. https://doi.org/10.3139/ 104.112107.

Erikstad, S.O. (2017). Merging physics, big data analytics and simulation for the next-generation digital twins. pp. 140–150. *In*: Bertram, V. (ed.). *HIPER'17: 11th Symposium on High-Performance Marine Vehicles*, Zevenwacht, 11– 13 September 2017, Technical University Hamburg-Harburg, Hamburg.

Grieves, M. and Vickers, J. (2017). Digital twin: Mitigating unpredictable, undesirable emergent behavior in complex systems. pp. 85–113. *In*: Kahlen, F.-J., Flumerfelt, S. and Alves, A. (eds.). *Transdisciplinary Perspectives on Complex Systems: New Findings and Approaches*. Switzerland: Springer International Publishing. https://doi.org/10.1007/978-3-319-38756-7_4.

Hardin, L.E. (2003). Problem-solving concepts and theories. *J. Vet. Med. Educ.*, 30(3): 226–229.

Hayes, A.W. (2005). The Precautionary Principle. *Archives of Industrial Hygiene and Toxicology*, 56(2): 161–166. https://www.researchgate.net/publication/7774245_The_Precautionary_Principle.

Jones, D.E. et al. (2019). Early stage digital twins for early stage engineering design. *Proceedings of the Design Society: International Conference on Engineering Design*, 1: 2557–2566. https://doi.org/10.1017/dsi.2019.262.

Morgan, J.P. (1996). *RiskMetrics™—Technical Document*. https://www.msci.com/documents/10199/5915b101-4206-4ba0-aee2-3449d5c7e95a.

Newell, A. and Simon, H. (1972). *Human Problem Solving*. Englewood Cliffs, NJ: Prentice Hall.

Passingham, R.E. and Rowe, J.B. (2016). *A Short Guide to Brain Imaging: The Neuroscience of Human Cognition*. Oxford University Press.

Polya, G. (2014). *How to Solve It. A New Aspect of Mathematical Method*. Princeton University Press.

Rescorla, M. (2020). The computational theory of mind. *In*: Zalta, E. (ed.). *The Stanford Encyclopedia of Philosophy*, (Spring 2020 Edn.). https://plato.stanford.edu/archives/spr2020/entries/computational-mind/.

Schleich, B. et al. (2017). Shaping the digital twin for design and production engineering. *CIRP Annals*, 66(1): 141–144. https://doi.org/10.1016/j.cirp.2017.04.040.

Schluse, M. and Rossmann, J. (2016). From simulation to experimentable digital twins: Simulation-based development and operation of complex technical systems. *In*: *ISSE 2016: 2016 International Symposium on Systems Engineering Edinburgh*, Scotland, George Hotel, 3–5 October 2016 proceedings papers, Edinburgh, United Kingdom,

10/3/2016–10/5/2016, IEEE, Piscataway, NJ, pp. 1–6. https://doi.org/10.1109/ SysEng.2016.7753162.

Schuh, G. and Blum, M. (2016). Design of a data structure for the order processing as a basis for data analytics methods. *In: Portland International Conference on Management of Engineering and Technology (PICMET)*, Honolulu, HI, USA, 4–8 September 2016, IEEE, pp. 2164–2169. https://doi.org/10.1109/ PICMET.2016.7806715.

Shafto, M. et al. (2010). Modeling, Simulation, Information Technology & Processing. *DRAFT Technology Roadmap Area*, Vol. 11, Washington, DC. Available at: https://www.nasa. gov/pdf/501321main_TA11-MSITP-DRAFT-Nov2010-A1.pdf.

Siedlak, D.J.L. et al. (2018). A digital thread approach to support manufacturing-influenced conceptual aircraft design. *Research in Engineering Design*, 29(2): 285–308. https://doi. org/10.1007/s00163-017-0269-0.

Smith, R.C. (2014). *Uncertainty Quantification: Theory, Implementation, and Applications*. SIAM, Philadelphia, PA.

Stark, R. and Damerau, T. (2019). Digital twin. pp. 1–8. *In*: Chatti, S. and Tolio, T. (eds.). *CIRP Encyclopedia of Production Engineering*, Vol. 66. Berlin: Springer Berlin Heidelberg. Heidelberg. https://doi.org/10. 1007/978-3-642-35950-7_16870-1.

Tao, F. et al. (2019). Digital twin-driven product design framework. *International Journal of Production Research*, 57(12): 3935–3953. https://doi.org/10.1080/ 00207543.2018.1443229.

Trauer, J., Schweigert-Recksiek, S., Engel, C., Spreitzer, K. and Zimmermann, M. (2020). What is a digital twin? – Definitions and insights from an industrial case study in technical product development. *International Design Conference – Design 2020*. https://doi.org/ 10.1017/dsd.2020.15.

von Neumann, J. and Morgenstern, O. (1953). *Theory of Games and Economic Behavior*. Princeton University Press.

# 17

# Application Examples
## Lesson Learned from the
## Case Studies*

Chapter 17 presents examples of three different disasters that have occurred over the past several decades, and documentation on the analysis of their course and potential causes is widely available. The first example concerns the collapse of a bridge on 1 August 2007, a stationary system that is part of the transportation infrastructure. In this case, the cause of the catastrophe was not unpredictable forces of nature, but the lack of a comprehensive analysis of the entire complex system and all processes taking place in it, both in terms of transport, repair, and logistics. The second example applies to the Three Mile Island nuclear accident on 28 March 1979. This catastrophe inspired Charls Perrow to develop the Theory of Normal Accidents, one of the conclusions of which is that human cognitive limits are important in causing system accidents. In the third example, an analysis was performed of the damage from the Great East Japan Earthquake and Tsunami on 11 March 2011. This example shows that classical probabilistic methods used in the assessment of extreme events of the natural disaster type cannot be the basis for decision-making in the of high-impact hazards. However, in these situations, methods based on the precautionary and resilient principles should be used.

* https://orcid.org/0000-0002-2630-3507.

## 17.1 Case Study I: Collapse of I-35W Highway Bridge Minneapolis, Minnesota, 1 August 2007

(Based on Highway Accident Report NTSB/HAR-08/03. Washington, DC. https://www.ntsb.gov/investigations/AccidentReports/Reports/HAR0803.pdf ).

### Accident Brief Description

At 6:05 p.m. 1 August 2007, the eight-lane, 1,907-foot-long I-35W highway bridge over the Mississippi River in Minneapolis, Minnesota, experienced a catastrophic failure in the main span of the deck truss. Consequently, 1,000 feet of the deck truss collapsed, with about 456 feet of the main span falling into the river. As a result of the bridge collapse, 13 people died, and 145 people were injured. Repaving Roadway work was underway on the I-35W bridge at the time of the collapse. On the day of the collapse, roadway work was underway on the I-35W bridge, and four of the eight travel lanes were closed to traffic. A few hours earlier, construction equipment and construction aggregates (sand and gravel for making concrete) had been delivered and positioned in the two closed inside lanes. The equipment and aggregates, which were being staged for a concrete pour of the southbound lanes that was to begin about 7:00 p.m., were positioned toward the south end of the deck truss portion of the bridge centre section and had been in place by about 2:30 p.m. Around 6:05 p.m., a video camera with active motion monitoring installed west of the I-35W bridge, recorded part of the collapse sequence. The video showed the middle span of the bridge separating from the rest of the bridge and flowing into the river. A total of 111 vehicles on the bridge at the time of its collapse were documented, 25 of which were construction vehicles. One of the 86 non-construction vehicles was a school bus carrying 63 pupils and a driver. After the collapse, 17 vehicles were found in the river or on the submerged part of the bridge deck.

### Emergency Response

As soon as the bridge collapsed, Minnesota State Patrol dispatchers were notified of the accident by cell phone through the 911 system. Some of the first to become involved in the rescue effort were about 100 citizens (including construction workers) who were near the area when the collapse occurred. Within the first hour of the collapse, 12 public safety agencies responded with 28 watercrafts to assist with river rescue operations. About 7:27 p.m., it was decided to change the water operations from rescue mode to recovery mode.

**Probable Cause**

Between the time of its opening in 1967 and the accident, the I-35W bridge underwent three major modification projects, two of which increased the dead load (i.e., the static load imposed by the weight of materials that make up the bridge structure itself) on the structure, namely:

- The 1977 construction plan for renovation of the bridge involved milling the bridge deck surface to a depth of 1/4 inch and adding a wearing course of 2 inches of low-slump concrete.
- Work done on the bridge in 1998 involved replacing the median barrier, upgrading outside concrete traffic railings, improving drainage, repairing the concrete slab and piers, retrofitting cross girders, replacing bolts, and installing an anti-icing system.
- Renovation started in June 2007 consisting in the removal of the concrete wearing course to a depth of 2 inches and adding a new 2-inch-thick concrete overlay.

The factors that were considered potentially causal or contributory to the accident:

- Insufficient bridge design firm quality control procedures for designing bridges, and insufficient procedures for reviewing and approving bridge design plans and calculations.
- Lack of guidance for bridge owners regarding the placement of construction loads on bridges during repair or maintenance activities.
- Exclusion of gusset plates in bridge load rating guidance.
- Lack of inspection guidance for conditions of gusset plate distortion.
- Inadequate use of technologies for accurately assessing the condition of gusset plates on deck truss bridges.

The National Transportation Safety Board determines that the probable cause of the collapse of the I-35W bridge in Minneapolis was the inadequate load capacity, due to a design error of the gusset plates at the U10 nodes, which failed under a combination of:

- substantial increases in the weight of the bridge, which resulted from previous bridge modifications, and
- the traffic and concentrated construction load on the bridge on the day of the collapse.

Contributing to the accident was the generally accepted practice of giving inadequate attention to gusset plates during inspections for conditions of distortion, such as bowing, and of excluding gusset plates in load rating analyses.

**Recommendations**

To improve the reliability of trussed bridges, it is essential to:

a) Develop and implement a bridge design quality assurance and control programme which includes procedures to detect and correct bridge design errors before the design plans are made final; and provides a means for verifying that the appropriate and accurate design calculations have been performed, as well as the specifications for the load-carrying members were adequate regarding the expected service loads of the structure.

b) Require bridge owners to determine where visual inspections may fail to detect corrosion and fatigue cracks in the gusset plate and where appropriate nondestructive evaluation technologies shall be used to assess the condition of the gusset plate.

c) Develop specifications and guidelines for use by bridge owners to ensure that construction loads and stored raw materials placed on a structure during construction or maintenance projects do not overload the structural elements or their connections.

d) Complete the Manual for Bridge Evaluation by including guidelines for assessing the load capacity of new bridges before they are put into service.

e) Modify the approved bridge inspector training as follows:

- update the training courses of the National Highway Institute to address inspection techniques and conditions specific to gusset plates, emphasizing issues associated with gusset plate distortion as well as the use of non-destructive evaluation at locations where visual inspections may not be sufficient to quantify parameters such as material loss due to corrosion; and

- include changes to reference materials, such as the Bridge Inspector's Reference Manual, and refer to any newly developed gusset plate condition ratings in the American Association of State Highway and Transportation Officials commonly recognized structural elements.

**Lesson Learned from the Case Study I**

In the case study I the object of the analysis is a stationary engineered system which is an important part of the transport infrastructure. It is, therefore, a typical object of interest for reliability engineering. Although each bridge is slightly different from the others, due to the need to adapt it to the local terrain conditions, the degree of similarity with such a high repeatability (there are over 50,000 bridges in Texas alone!) is very high. The choice of this bridge for analysis was dictated by the fact that the main cause of this accident was not a typical one (natural disaster such as flood, hurricane, or earthquake) but an overlap of various internal (degradation processes such as corrosion and material fatigue) and external (repair work on the bridge) factors. Moreover, an important criterion in the selection of this case was the availability of documentation of the disaster temporary course, thanks to the recording of the collapse sequence from the motion-activated surveillance video camera at the Lower St. Anthony Falls Lock and Dam, just west of the I-35W bridge.

The main conclusions of this case are as follows:

- Stationary engineered systems are major components of critical infrastructures (ICs) and are found in huge numbers on a global scale, so their reliability is a key condition for ensuring the continuity of IC operations.

- A significant part of these systems was designed and built many years ago (e.g., the I-35W bridge was 40 years old at the time of the accident), i.e., when the calculation methods were not as accurate as nowadays and the expected operating loads were much lower.

- Modifications of the above-mentioned systems undertaken to adapt them to higher loads were limited by their structure and local conditions, and generally resulted in a significant increase in their weight. This means that a significant part of the increased load capacity of the system was reduced by a significant increase in the weight of the system, and therefore the positive effect of the modification was negligible in practice.

- The ever-increasing intensity of ICs use, and economic pressure mean that repair and modification work is carried out without the systems being completely out of service, reducing only partially the intensity of operation.

- Logistical aspects, such as the organization of the supply of materials and their storage at the site of renovation and/or modification, further increase the risk of temporary overloading of the system, which can lead to disaster under unfavorable circumstances.

- Decisions on the future operation of these systems and how to renovate or upgrade them are based on incomplete and uncertain knowledge and are burdened with human factor imperfections (cognitive aspect), which further increases the level of risk.

- Due to the above, despite the seemingly low complexity of the systems in question, the problem of ensuring their reliability is a complex and difficult one. It is therefore necessary to develop and implement a comprehensive reliability assessment method for the whole life cycle of these systems, considering all factors influencing their lifetime, with a particular focus on safety and cognitive aspects (Moura Beer et al., 2016).

## 17.2 Case Study II: The Three Mile Island Nuclear Accident, 28 March 1979

(Based on Analysis of Three Mile Island – Unit 2 Accident; NSAC-80-1; NSAC-1 Revised March 1980; Prepared by the Nuclear Safety Analysis Center, https://www.nrc.gov/docs/ML1205/ML12055A064.pdf).

### Accident Brief Description

On 28 March 1979, occurred an accident of a nuclear power plant at the Three Mile Island (TMI) site 10 miles southeast of Harrisburg, Pennsylvania. Failure of the cooling system of the Unit 2 nuclear reactor led to overheating and partial melting of the pressurized-water reactor's uranium core which has resulted in release of radioactive gas and contaminated water. The accident had many root causes, related both to technical malfunction in the condensation system and human error. Three days after the accident, the official issued an advisory to evacuate pregnant women and preschool children living within a 5-mile radius of Three Mile Island, raising fears an explosion and dispersal of radioactivity among residences. As a result, about 140,000 people evacuated the area in panic. In addition, this accident contributed to public concern over the dangers of nuclear power and slowed construction of other reactors in the US.

The course of the incident can be presented in three stages:

a) The Three Mile Island Generating Station (TMIGS) had two pressurized water reactors. The accident in the Unit 2 reactor (with a power of about 960 MWe) began when the plant's main feed-water pumps in the secondary non-nuclear cooling system failed. In response to this malfunction, the auxiliary feed-water pumps kicked in, however, the water did not reach the steam generator because the outlet valves were closed. However, the fact that the valve was closed was only discovered about eight minutes after the accident. In response to the increase in temperature and pressure in the primary system, the pressurizer relief valve automatically opened in the line between the pressurizer and the quench tank. After the nuclear reactor automatically shut down, pressure in the reactor dropped. The valve, which seemed to be closed after the pressure dropped below the set-point for closure, failed to re-close. The valve was left open for 140 minutes and leaking 80 tonnes of primary coolant water from the quench tank. At that time more than 100 warnings appeared in the control room causing great panic.

b) Pressure-drop in the reactor caused the Emergency Core Cooling System (ECCS) to pour water into the reactor system at a speed of 4 ton/min. As water and steam escaped through the relief valves, the coolant water fell into the pressurizer, raising the water level in it. Not knowing the valve was jammed and the display was false, operators manually shut down the ECCS. The temperature rose and then steam was generated in the original reactor cooling system. Pumping the steam-water mixture caused the reactor cooling pumps to vibrate, and the operators turned off the pumps to stop the strong vibrations. When the reactor cooling water boiled, the top of the reactor core was exposed.

c) Coolant water stopped flowing into the pressurizer, the temperature in the primary system reached 2,000 degrees Celsius, and approximately 45% of the reactor core was melted. The coated tubing and water reacted to produce hydrogen gas, and 10 hours later, there was a hydrogen explosion. Water from the pressure relief valves overfilled the quench tank and flooded the containment vessel floor. The floor pump sent the radioactive coolant to the ventilation system of the auxiliary building, where the radioactivity escaped outside. Eventually, cooling water was added and the reactor which then began to cool with natural circulation, however, hydrogen and radioactive gas was still being produced and about 10 million curies of radioactive gas were released into the atmosphere because of this incident.

**Probable Cause**

The chain of causes that led to this incident can be attributed to the individual activities in the extended resilience matrix (see Section 16.4), namely:

No. 2 – Identifying.

The process of identifying sources of risk did not consider the impact of operators' competence on the reliability and safety of system operation. The operators of the Three Mile Island Nuclear Power Plant were not employees of the utilities firm. The company had contracted out the plant operations, however, the contract operators lacked adequate knowledge about nuclear reactors and thermal phenomena. They were barely trained in emergency situations.

No. 4 – Monitoring and No. 5 – Detecting.

Problems with the control room's failure information system caused inadequate response of operators to emergency situations. The warning sign blocked the view of the LED indicating that the auxiliary water supply valve was closed, and in addition, when the valve was closed, the green LED was lit. The instruments in the control room (over 1200

LEDs on the control panel) did not have the same LED colour – some of the LEDs were green and some were red under unusual conditions. The instruments only showed that a CLOSED signal was sent to the relief valve without indicating the actual valve position. The LEDs were not designed to warn of a malfunction and operators did not know that the relief valve was open.

Operators generally try to prevent the pressurizer from being filled as they will not be able to control the pressure in the water-filled pressurizer. Note that the water level in the pressure gauge rises with high pressure injection pumps that push replacement water into the reactor system and cooling water enters the pressurizer while water and steam escape through open relief valves. The production of gas from the heated reactor core also raises the water level in the pressurizer. Operators have incorrectly assessed that the water level in the pressurizer increases in such situations. The water level indicator did not show the actual water level in the pressurizer.

No. 6 – Response.

As a result of a poorly designed monitoring system and incorrect interpretation of the signals during the analyzed incident, instead of reducing the negative consequences of the accident, operators' actions increased them even more. An example of such a situation is the fact that unaware of the stuck valve and the false indicator readings, operators manually turned off the Emergency Core Cooling System. The consequence of this malfunctioning was an uncontrolled increase in temperature and the rapid formation of steam in the original reactor cooling system. Finally, the reactor cooling water boiled away, and the top of the reactor core was exposed.

No. 7 – Repairing.

The malfunctioning relief valve of the pressurizer already had repeated problems and was unreliable. Despite the obvious problem, the plant management instructed the operators to continue operating the system without replacing the valves with more reliable ones. As a result of such a policy (lack of reliability assurance), system maintenance did not meet the basic safety requirements.

No. 8 – Learning.

Unfortunately, the system for operating the TMI nuclear power plant did not provide for the possibility to learn from incidents and minor accidents. As a result, a catastrophe occurred, resulting in great economic and image damage, and about 140,000 people were evacuated from the affected area in panic. Only after the disaster the U.S. President ordered a full investigation of the incident, which resulted in 52 recommendations including safety standards, safety study, safety design, operating control, disaster prevention and safety research. In addition, the local government reviewed its nuclear disaster prevention plan.

No. 9 – Evolving.

The result of the error learning process was a modification of the system and its management procedures, namely:

- installing highly reliable equipment and maintain a backup system so that it functions properly when needed,
- designing safety devices that has easy-to-read indicators accommodating human cognitive constraints,
- developing safety systems for preventing inadvertent control inputs and erroneous operations,
- preventing outsourcing from degrading operation efficiency and reliability.

No. 10 – Anticipation.

Safety devices in nuclear power plants are designed to handle certain nuclear accidents. The chain of events in a nuclear accident on Three Mile Island went well beyond the scope of the assumed accidents and no one thought about how to deal with such cases. However, prior to this accident, similar accidents were reported at a Swiss power station in August 1974, at Oconee Nuclear Station in the US in June 1975 and at Davis-Besse Nuclear Power Station in the US in September 1977, and their reports containing an analysis and assessment of the events were available. Business managers and, particularly, the management of utilities should have carefully analysd these accident reports and taken preventive measures.

**Lesson Learned from the Case Study II**

The object of the analysis in the case study II is a complex production system which is a key part of the energy infrastructure. Due to the much higher degree of complexity of this system in comparison with the bridge construction (case study I) and the significant impact of the human factor (operators) on the system, the analysis of causes of this accident was far more difficult. Moreover, unlike the I-35W bridge I-35W, the reactor's design was new and the unit itself was only in operation for a few months.

The Three Mile Island accident inspired Charles Perrow to write the book titled *Normal Accident Theory* (Perrow, 1984), in which he made the thesis that an accident occurs, resulting from an unanticipated interaction of multiple failures in any complex system. TMI-2 was for him an example of this type of accident because it was "unexpected, incomprehensible, uncontrollable and unavoidable". 'Normal' accidents are so-called by Perrow because such accidents are inevitable in extremely complex systems, in which multiple failures interact with each other, despite efforts to avoid them, whereas, events which appear trivial initially cascade and multiply unpredictably, creating a much larger catastrophic event. In his next book titled *The Next Catastrophe: Reducing Our Vulnerabilities to Natural, Industrial, and Terrorist Disasters* (Perrow, 2007), has refined and extended his Normal Accident Theory. Perrow pointed out the dangers of people becoming dependent on tightly coupled, interactively complex, difficult to control, under-regulated, and vulnerable technologies. He explained socio-technical and political processes concerning disaster vulnerabilities in a simple and clear way on examples of catastrophes from previous years (e.g., catastrophe 9/11 from 2001, Hurricane Katrina from 2005 and cases of Internet terrorism). Perrow postulates that there are three general causes of disaster: natural causes, organizational causes, and deliberate causes, and there are four categories of failure: executive failure, mundane organizational failure, technological failure, regulatory failure.

Perrow recommends the following as a remedy to prevent unexpected accidents and catastrophes:

- building redundancy into important systems, wherever technically and economically justified,
- decentralization of the management of complex systems instead of their fragmentation,
- limiting deregulation and self-regulation only to justified cases which do not endanger the safety of people and the environment.

From the perspective of more than 40 years since this accident, it is difficult to agree with the opinion that accidents are inevitable in complex systems, because the Cyber-Physical-Social Systems currently in use are much more complex than a nuclear power plant TMI-2 and have a very high level of reliability and safety. This can be supported by the fact that Generator 1 in the TMI power plant (with the same design as generator 2), following the changes recommended in the accident report, was still operating successfully until September 2019. However, the author of the Normal Accident Theory was right to say that human cognitive limits are important in causing system accidents.

## 17.3 Case Study III: Damage from the Great East Japan Earthquake and Tsunami, 11 March 2011

(Based on Fukushima Nuclear Accident Analysis Report, 20 June 2012 Tokyo Electric Power Company, Inc., https://www.tepco.co.jp/en/press/corp-com/release/betu12_e/images/ 120620e0104.pdf).

### Accident Brief Description

On 11 March 2011, an earthquake of magnitude 9.0 occurred in the Pacific, just off the coast of Tohoku, Japan. The combined earthquake and resulting giant tsunami caused enormous damage in eastern Japan. Its run-up height reached over 39 m, and over 24,000 people were reported as dead or missing. The number of temporary refugees exceeded 350,000 at one time. Moreover, serious accidents at the Fukushima Nuclear Power Plants (NPP) No. 1 of Tokyo Electric Power Co. (TEPCO) were caused by the tsunami.

The main characteristics of the Great East Japan Earthquake were as follows:

- The earthquake was the biggest measured earthquake in Japan. Maximum recorded acceleration of the ground motion reached 3,000 gal (30 m/s$^2$) in Kurihara-city (Miyagi prefecture), and shaking lasted for nearly 6 min.

- Numerous large aftershocks followed the main event until 19 April (over 420 with magnitude of 5.0 to 7.1). These designate that the phenomena were a series of large- scale crustal motions, with ongoing consequences.

- The leading period of the main earthquake was 0.2 to 1.0 s, which corresponds to the frequency 1 to 5 Hz.

- Shortly after the major earthquake the first wave arrived at cities near the failure fault, such as Miyako, Ofunato, and Kamaishi in Iwate Prefecture. The highest waves of nearly 13 m high reached the nearest coastlines 30 to 40 min after the earthquake occurred.

Wave heights and run-up elevations of tsunami are not the same; wave height is the distance between the mean sea level and the crest of the tsunami wave in the sea, while run-up elevation refers to the maximum height of the land where the tsunami reaches. For example, in Rikuzen-Takata-city, the wave height was over 15 m, with the highest run-up elevation being about 39 m. Furthermore, tsunami waves extended up along a river and flooded over river dykes in the middle and upper reaches.

The earthquake resulted in huge damage, namely (Mimura et al., 2011):

- Human casualties include 14,508 dead and 11,452 missing people.

- The number of completely collapsed and washed-out houses amounts to 76,000, and the number of those with partial damage was over 244,000. Following the earthquake, 345 fires occurred, including cases where the tsunami triggered the fire.

- Infrastructure damage was also very wide-spread; reported damaged included 3,546 areas along roads, 71 bridges and 26 parts of the railway system. However, thanks to efforts for disaster preparedness, the hundreds of trains in operation were able to make emergency stops safely, without any deaths or serious injuries.
- Critical infrastructure such as electricity, water supply, sewage systems, and gas lines, was also damaged. Though such services were soon restored in most of the damaged areas, the coastal areas most heavily damaged in northeast Japan are for long time without these services at the time of writing.
- The damage costs were estimated at almost 22 trillion yen ($188 billion), but the newest calculations increase the total costs up to 80 trillion yen ($736 billion) in over 40 years.
- The largest consequence of the earthquake and tsunami was the accident at the Fukushima Nuclear Power Plant (NPP) No. 1. Many serious problems have resulted, including nuclear contamination of soils and sea water, and evacuation of many residents in the surrounding areas, initially 20 km and then subsequently more distant from the NPP. Because the physical impacts of the earthquake and tsunami were combined with health, psychological, and social problems caused by the Fukushima NPP accident, the situation became especially serious and complicated.

### The Accident at the Fukushima Nuclear Power Plant No. 1

Construction of the first nuclear reactor in Fukushima started in 1967, and six reactors were built over a period of about 10 years. The generation capacity of the Fukushima No.1 was approximately 4,700 MW in total. Fukushima NPP No. 2 is located 10 km south of Fukushima No. 1. These two NPPs are all situated on the coast, with their own ports for transportation of nuclear fuels and cooling water uptake. The Fukushima coast, therefore, is Japan's big electricity supply base.

When on 11 March 2011 at 2:46 p.m., an earthquake of magnitude around 9.0 was detected, safety measures were immediately initiated. Nuclear reactors 1 to 3 made emergency shutdowns successfully, while the other reactors 4 to 6 were not operating because of regular inspection. As the electric feeder lines to Fukushima NPP No. 1 were damaged by the earthquake, emergency generators started to power the emergency cooling system. However, the first wave of the tsunami arrived at 3:27 p.m., and the second wave 8 min later, inundating the buildings housing the nuclear reactors and generators with water 4 to 5 m deep. As a result, the emergency generators stopped working, and reactors 1 to 5 lost electricity for cooling systems. After one day, cooling water levels were lowered in reactors 1 to 3, resulting in damage and partial melting of nuclear fuel rods. To protect the nuclear reactors, pressure was lowered by releasing water vapour out of the reactors. This diffused radioactive matter into the atmosphere.

**Probable Cause**

The course of the disaster can also be analysed from the perspective of the activities listed in the extended resilience matrix (see Section 16.4) as follows:

No. 2 – Identifying and preparing.

Tsunami hazard maps have been prepared for each affected area. These maps showed, based on scientific research, the estimated inundation areas of tsunamis and river floods, as well as landslide risk areas. They also promoted soft measures in local disaster prevention workshops, such as discussions how to find safe evacuation places and routes. Loudspeakers were installed in each town to inform about emergency cautions. These prevention activities had a great effect, even in face of the gigantic tsunami. For example, in Kamaishi nearly 3,000 children in elementary and middle high schools were evacuated safely. It is remarkable that this action was taken before local authorities issued evacuation alerts.

No. 3 – Protecting.

The Tohoku coast which was heavily damaged by the 11 March event had developed the most advanced anti-earthquake/tsunami system in the world. This region has suffered from large tsunamis in the past: in 1886 (Meiji Snriku Tsunami), 1933 (Showa Sanriku Tsunami), 1960 (Chilean Earthquake Tsunami), and 1968 (Off Tokachi Tsunami). Based on these disasters, large breakwaters were constructed at the mouths of Ofunato, Kamaishi, Miyako, and Kuji bays, to protect the cities located in the inner bay areas against tsunamis. For example, Kamaishi Bay has a breakwater with a total length of 2 km and maximum depth below water 63 m, while the height above mean sea level is 8 m. This breakwater was completed in 2009, using 700 million m$^3$ of concrete. This breakwater is the deepest breakwater in the world. Taro district, in Miyako City, was surrounded by high coastal dykes called "Great Walls" with 10 m high and 10 km in total length.

The Great East Japan Tsunami flew over these structures, attacking cities and causing huge casualties. The bay-mouth breakwaters were destroyed in Kamaishi and Ofunato. The great coastal dykes in Taro could not prevent overflows and were destroyed as well, but evidence confirms that the coastal structures reduced the damage to some extent.

No. 4 – Monitoring and No. 5 – Detecting.

Many tidal stations recorded high water levels of around 8 m in these areas, before they were destroyed by the tsunami, and no more data were recorded. One exception was Onagawa NPP of Tohoku Electric Power Co that recorded temporal variations of sea surface elevation at the coast during the catastrophe. In total, nine of the 10 tidal stations could not send their data after the first or second tsunami wave, or immediately after the earthquake. So, the whole region had lost its tsunami monitoring capability for a certain time. The tsunami damaged several electronic personal alarm dosimeters (PADs). The surviving dosimeters could not be recharged because of the electrical blackout at the site. Moreover, after the accident TEPCO could not operate the Internal Exposure Monitoring that were in the affected plant because of the increase of the background radiation level.

No. 6 – Responding and No. 7 – Repairing.

The big challenge was the interruption to transportation in the coastal areas. Ports, railways, and roads could not be used due to heavy damage; thus some places could be accessed only by helicopter. In effect, air transportation was the only way to bring water, food, and other life necessities to islands for 2 weeks. Disaster wreckage was another big problem. The earthquake and tsunami generated 25 million tonnes of wreckage in only the three prefectures of Iwate, Miyagi, and Fukushima. It was practically impossible to take the wreckage to outer areas. Stockpiling a mixture of wreckages may cause contamination of soils, sandy beaches, and surface and ground waters. Thus, disasters and their removal are closely linked to environmental problems.

No. 8 – Learning, No. 9 – Evolving, and No. 10 – Anticipating.

The most significant feature of the Great East Japan Earthquake and Tsunami is in its *complexity*. We should regard it as not a simple composite disaster but an enormous composite disaster. The combination of an earthquake of magnitude 9.0 and a giant tsunami was unprecedented in Japan in the last 1,000 years. Additionally, the combination of these natural events and the nuclear accident makes this disaster with serious and widespread damage. Japan has for a long time been increasing the preparedness against earthquakes and tsunamis. However, huge damage occurred, because the catastrophe was beyond past experiences and predicted expectation. This event suggests that we should not only implement the traditional disaster prevention based on basic knowledge, but also consider the possibility of *maximum potential hazards*.

To save lives and resources of a in the face of such a giant tsunami, there are no ways other than quick evacuation and rescue. At the same time, we can protect human lives and socioeconomic activities against more frequent and lower magnitude tsunamis by using a combination of hard and soft measures. An idea emerging in the discussion for reconstruction of the northeast Japan is division of living and working places. People should live in higher places, while they work on the coastal areas because low-lying coastal areas are still effective and important for ports, fishery ports and commercial activities. Therefore, coastal dykes, coastal forests, and other facilities should be reconstructed so that they can protect the economically important coastal areas against using conventional hazard assumptions and classical calculation methods. We need to establish flexible and disaster prevention systems with multilayered targets. At present, there were barriers among different fields to hamper information flows between disciplines. Single academic fields cannot resolve real, complex problems, therefore, *multi-disciplinary approaches* and cooperation are necessary to meet these challenges.

### Lesson Learned from the Case Study III

A *hundred-year wave* is a statistically predicted water wave whose height is reached or exceeded on average once every hundred years for a given location. It has a 63% probability of reaching this wave height at least once during a 100-year period. As a forecast of the most extreme wave that can be expected to occur in a given water body, a 100-year wave is a factor commonly considered by designers of oil platforms and other offshore structures.

Based on such probabilistic assumptions, Japanese experts in 2006 calculated that the probability of exceeding the 6-metre wave height value in the event of a tsunami in the Fukushima area is less than 0.01 (1%) in the next 50 years (i.e., until 2056). In fact, after just five years, a tsunami occurred, resulting in a wave height of 14 m. This demonstrates the inadequacy of the probabilistic methods used and the need to reach, if possible, to data from much longer time frames, such as 1,000 years. In the case of the catastrophe under analysis, this was possible, because historical data testified that as early as 869 an earthquake with a magnitude of about 8.6 occurred in Sanriku, and in its aftermath a tsunami caused sea waves to break 4 km inland. On the other hand, in 1611, an earthquake with a magnitude of about 8.1 caused waves as high as 20 metres (Pate-Cornell, 2012).

The basic conclusion of the above analysis may be that classical probabilistic methods used in the assessment of extreme events of the natural disaster type cannot be the basis for decision-making in the case of high-impact hazards. Especially not recommended are methods of time series analysis, in which historical data are weighted according to the principle 'the older the less relevant'. Much more reliable are methods based on extreme value theory, using only the maximum or minimum values of the parameter under study from specific time intervals for statistics. However, under conditions of deep uncertainty, as a rule, such data are not available, so it is then recommended to use the concept of cognitive reliability, which is presented in Chapter 16 of this book. Although this method does not provide a quantitative assessment of the risk associated with an adverse event of high harmfulness and low probability of occurrence, but in accordance with the principle of precaution and resilience, it protects in the most reasonable and practicable way possible against the negative consequences of such an event.

## 17.4. References

https://www.ntsb.gov/investigations/AccidentReports/Reports/HAR0803.pdf.

https://www.nrc.gov/docs/ML1205/ML12055A064.pdf.

https://www.tepco.co.jp/en/press/corp-com/release/betu12_e/images/    120620e0104. pdf.

Mimura, N. et al. (2011). Damage from the great East Japan Earthquake and Tsunami: A quick report. *Mitigation and Adaptation Strategies for Global Change*, 16(7): 803–818.

Moura, R., Beer, M., Patelli, E., Lewis, J. and Knoll, F. (2016). Learning from major accidents to improve system design. *Safety Science*, 84: 37–45.

Pate-Cornell, E. (2012). On 'Black Swans' and 'Perfect Storms': Risk analysis and management when Statistics are not enough. *Risk Analysis*, 32(11): 1823–1833.

Perrow, Ch. (1984). *Normal Accidents: Living with High-Risk Technologies.* New York: Basic Books.

Perrow, Ch. (2007). *The Next Catastrophe: Reducing Our Vulnerabilities to Natural, Industrial, and Terrorist Disasters.* Princeton University Press.

# 18

# Summary and Concluding Considerations[*]

In line with current trends in business and engineering, represented, among other things, by the concept of *Industry 5.0*, the main motivation for writing this book was the author's belief in the need to combine two different approaches, namely human-centred and technology-exposing. In addition to the many obvious differences between the two approaches, due to both tradition and the difference between the evaluation criteria, uncertainty and the way in which risk is perceived and evaluated appear to be the primary factors hindering the effective solution to this problem. It is especially true for complex systems of the so-called Cyber-Physical-Social nature, which in the modern world have a global scope and significantly affect almost all areas of our lives. Since these issues are very broad, this paper focuses only on selected aspects that relate to the operation and maintenance of these systems. These areas of knowledge are dealt with by Systems Engineering and, particularly, Reliability Engineering, Safety Engineering, Security Engineering, Resilience Engineering, and Dependability Engineering. Of the above, only Reliability Engineering has its own theory, which is mathematically formalized, based on the probability theory and mathematical statistics (e.g., Barlow and Proschan, 1965; 1975). Other attribute-oriented systems engineering is in constant development, have no established and universally accepted scopes, and even about their definition there is no clear consensus among specialists. This has resulted in various attempts to expand their scope into areas of other attribute engineering. It is particularly true of Safety Engineering, which in its new versions known as Safety II (e.g., Hollnagel, 2018), and especially Safety III (e.g., Leveson, 2020), also include the scope of Security Engineering and Resilience Engineering. This is evidenced, among other

[*] https://orcid.org/0000-0002-2630-3507.

things, by the following excerpt from the paper written by Nancy Leveson (2020):

**Safety-III is based on Systems Theory. It spans the entire lifecycle but puts particular focus on designing safety in from the very beginning of system concept definition. Resilient systems are created not by simply focusing on human operators but instead by carefully designing the system to prevent and control hazards as much as possible, including the operator as a critical component of that design. In addition, if emergencies do arise that the system is not designed to handle, the system is designed so that human operators can be successful in handling it and that the tools exist for humans to be resilient in the case of emergencies. Safety-III recognizes that change and adaptation to change are both inevitable and healthy in the system's lifetime. Planned changes must be carefully analysed to ensure that they do not increase risk nor introduce new hazards.**

Thus, according to this concept, Safety III integrates not only elements of Security and Resilience Engineering, but also Risk Analysis, with a special focus on the human factor. Aven (2020), on the other hand, believes that the field of Resilience Engineering is closely related to the area of Risk Analysis, writing, among other things:

**The resilience field arose as a supplement to the traditional probabilistic risk assessment approach, which has strong limitations in analyzing many types of real-life systems, in particular complex systems, which are characterized by large uncertainties and potential for surprises. By strengthening the resilience of the system, the safety is enhanced without a need to perform risk calculations. … The attractiveness of the resilience approach is that we do not need to know what type of events – hazards and threats – can occur and to express their probabilities as needed in traditional risk assessment.**

Today's practice in both manufacturing and services needs an integrated system that considers all the aspects mentioned above. This can be demonstrated by the *Industry 5.0* concept, which is based on three basic pillars (EU, 2021):

- Human-centric approach, which places human needs at the heart of the production process, asking what technology can do for workers and how can it be useful.

- Sustainability, which focuses on reuse, repurpose, and recycle of natural resources and to reduce waste and its environmental impact.

- Resilience, which implies an introduction of robustness in industrial production. This robustness provides support through flexible processes and adaptable production capacities, especially when a crisis occurs.

It follows from the above that there is currently a need to integrate all activities aimed at achieving the mentioned goals, which is also to be served by the concept of Cognitive Systems Engineering, presented in this book. The foundations of this concept are the definitions and pre-paradigm proposed by the author (Chapter 11):

Cognitive Dependability is the ability of human-centric Cyber-Physical-Social Systems to perform trustworthy in a variety of situations, including conditions of deep uncertainty.

Cognitive Dependability Engineering is a generalized version of the Dependability Engineering targeting complex human-centric Cyber-Physical-Social Systems which should perform trustworthy in a variety of situations, including conditions of deep uncertainty.

**The concept of Cognitive Dependability Engineering is based on cognitive processes leading to subjective, largely instinctive expert judgment, which in engineering practice may be supported by intuitive models and elicitation procedures.**

The book is structured into four parts that answer the following main questions:

- What concept to use to define the research objects of this work?
- What attributes of these objects are critical to achieving their goals, and what tools are best suited to do so?
- What methods should be used to study these objects and their behaviour in various situations, especially unusual and risky ones?
- How to solve the problems of ensuring the trustworthy operation of these facilities?

The purpose of Part I is to answer the first question, so it is devoted to an overview of the systems engineering development: starting with basics of System Science and Single Systems Engineering, through System of Systems Engineering to Cognitive Systems Engineering. The Cognitive Systems Engineering model was based on the concept of imperfect knowledge acquisition and management.

Answering the second question, Part II presents five main concepts of attribute-oriented System Engineering: from Reliability Engineering – derived from the idea of failure-free operation, through Safety Engineering and Security Engineering – based on the concept of effective protection; Resilience Engineering – built on the principle of process continuity; Dependable Computing – designed on the concept of fault-tolerant functioning; and ending with the new concept of Cognitive Dependability Engineering. It is a transdisciplinary concept for ensuring that complex Cyber-Physical-Social Systems perform in a trusted manner, both under normal and abnormal conditions.

Part III deals with modelling and simulation the operation of Cyber-Physical-Social Systems in a risky environment, as an answer to the question about the methodology to study this issue. It consists of three chapters covering the following topics: methodology of modelling and simulation used for complex systems; modelling of structures, processes, and disruptions for Cyber-Physical-Social Systems; as well as simulation of Cyber-Physical-Social Systems behaviour in a risky environment.

The last, Part IV, deals with the problems of managing risks in Cyber-Physical-Social Systems under deep uncertainty, answering the last of the questions. Uncertainty-oriented decision-making concepts are discussed first, followed by a proposed method of cognitive dependability based problem-solving in Cyber-Physical-Social Systems operating under deep uncertainty. The concept of a cognitive digital twin was introduced to support the process of solving complex problems by experts, and on this basis a framework for cognitive dependability based problem-solving in Cyber-Physical-Social Systems operating under deep uncertainty was developed. The possibilities and purposefulness of using this framework have been demonstrated on three practical examples of disasters that have happened in the past and have been thoroughly analysed, and the descriptions of reports from these studies have been made available on the Internet. Chapter 17 describes the lessons that can be drawn from these cases regarding the following disasters: collapse of I-35W Highway Bridge Minneapolis in Minnesota on 1 August 2007; the Three Mile Island Nuclear Accident on 28 March 1979; and damage from the Great East Japan Earthquake and Tsunami, 11 March 2011.

In conclusion, the proposed concepts of cognitive dependability and cognitive dependability engineering have allowed the development of a framework for cognitive dependability based problem-solving in Cyber-Physical-Social Systems operating under deep uncertainty. The author is aware that this framework is only a general concept for solving operational problems under deep uncertainty, and therefore requires further work, both theoretical and experimental. To this end, further research is planned to be carried out by an interdisciplinary team consisting primarily of specialists in engineering, psychology, and cybernetics. A platform connecting these activities could be the concept of *flexible thinking*, which Leonard Mlodinov formulates as follows (Mlodinov, 2018):

… the capacity to let go of comfortable ideas and become accustomed to ambiguity and contradiction; the capability to rise above conventional mind-sets and to reframe the questions we ask; the ability to abandon our ingrained assumptions and open ourselves to new paradigms; the propensity to rely on imagination as much as on logic and to generate and integrate a wide variety of ideas; and the willingness to experiment and be tolerant of failure. …

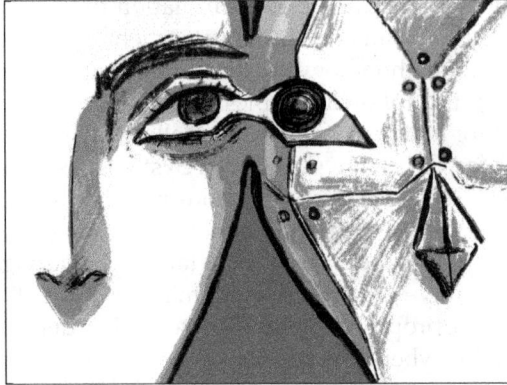

**Figure 18.1:** Towards emergent cognitive human-cyber intelligence (with permission of the author Weronika de Bończa Bukowska).

On this platform, an attempt can be made to create a system composed of two subsystems: a natural one based on cognitive expertise (human intelligence) and a cyber one resulting from the capabilities of a 'digital twin' (artificial intelligence). An abstract futuristic model of this concept might be Figure 18.1. It illustrates the cognitive process carried out by both subsystems in a holistic manner, as symbolized by a pair of eyes consisting of the human eyeball (the one with the pupil) and the eye-camera of the digital twin. Also, the other senses (organs and instruments) should form a coherent whole, to ensure the effective pursuit of a common goal of solving a specific problem. The rapid progress achieved in recent years in building ergonomic and reliable interfaces of the 'human brain – computer or robot' type (e.g., Wang, 2022) gives hope that this idea can be realized in practice in the not-too-far future.

# References

Aven, T. (2020). *The Science of Risk Analysis. Foundation and Practice*. London and New York: Routledge, Taylor, and Francis Group.

Barlow, R.E. and Proschan, F. (1965), *Mathematical Theory of Reliability*. New York, London, Sydney: John Wiley & Sons, Inc.

Barlow, R.E. and Proschan, F. (1975). *Statistical Theory of Reliability and Life Testing. Probability Models*. Holt, Rinehart, and Winston Inc.

EU. (2021). European Commission. Directorate General for Research and Innovation, *Industry 5.0—Towards a Sustainable, Human-Centric and Resilient European Industry*. Luxemburg: Publications Office of the European Union.

Hollnagel, E. (2018). *Safety-II in Practice*. UK: Routledge.

Leveson, N. (2020). *Safety III: A Systems Approach to Safety and Resilience*. MIT 7/1/2020. http://sunnyday.mit.edu/safety-3.pdf.

Mlodinov, L. (2018). *Elastic. Flexible Thinking in a Time of Change*. UK: Penguin

Wang, L. (2022). A futuristic perspective on human-centric assembly. *Journal of Manufacturing Systems*, 62: 199–201.

# Index

For Product Safety Concerns and Information please contact our EU
representative GPSR@taylorandfrancis.com
Taylor & Francis Verlag GmbH, Kaufingerstraße 24, 80331 München, Germany